THE TECHNOLOGY OF GOD

MARK HAWKINS

THE TECHNOLOGY OF GOD

THE THEORY OF EVERYTHING

Science, Religion & Law, LLC
An Arizona Company
www.sciencereligionlaw.org

ISBN: 978-0-9980368-1-6

CONTENTS

INTRODUCTION

This book is a theory of everything, including God.

Everything is technological. Evolution is technology. Mortals are technology, developed using evolution and DNA engineering. This theory is an upgrade over "God did it;" it shows why, not just how. It shows God as the first scientist, and an inventor of technology.

While God developed all technology in the past 14 billion years, he doesn't necessarily own it. The technology could exist independently. It is this insight which raises the possibility that God could give us everything, at least everything mortal, and still hide from us. It may tell us why he would want to.

Physics has a theory of everything, but it is limited to physics, and even then it only describes mortal matter, which makes up 20 percent of the universe.

Dark matter is eighty percent of the universe, and science knows virtually nothing about it.

Religion and science, combined, transcend mortal matter, and potentially explain the other eighty percent.

Technologically, there are four paradigms of matter. Mortal matter, made of atoms, is only one of them. We received data about the other three paradigms from eye witnesses. They described a person who looks like a mortal, who ate food, but who was indestructible, and descended from the sky. He was high tech. The problem is, the eye witnesses were religious, so we discount the data.

Atheistic evolutionary biology has a glaring weakness: It results in humans. Even if the theory were gloriously true, look at the result: We live eighty years, and we are unhappy a lot. On the one hand, we are high tech. A human is a wonderfully advanced piece of technology, however produced. On the other hand, we only live eighty years, and then our cells self-destruct. In a way, compared to billions of years we talk about, that is pathetic. And what do we look forward to? Oblivion. That is a rotten use of technology.

If there are four paradigms of planets or technology, as religion says, everything changes. We are going to evolve further. We are going to become permanent physical bodies, indestructible.

I have written in this book my theory of everything.

Science has weaknesses. Religion has weaknesses. Together, science and religion are transcendent. Combined with law, philosophy, and everything else, they reveal a zenith of society, nearing an end. If you don't like religion, forget about it, because you can. You will still be treated fairly in the overall system.

This world is programmed, scientifically and technologically, like a computer.

With or without God, the program will execute.

I use speculation as a tool. I use it as a weapon. I use it as entertainment. This book is highly speculative.

Here is a summary of the book:

God reproduces planets using a 3-D printer.

One reason God digitized the planet is that at the end of the world we need a spare to serve as part of heaven. Digitizing the planet for faster reproduction seemed like the obvious answer.

For God to move a mountain, or change water to wine, he would need the chemical structure and DNA digitized, and he would have to reproduce it by remote command. Hence the 3-D printer.

The point is that God is a scientist, not a magician.

Imagine the level of technology required to reproduce an exact replica of the planet earth in seven days. Of course God has computer technology; by all indications, our supercomputers are a poor mimicry of his technology. Whether he uses his technology to make digital copies of planets is another matter.

While the concept of reproducing planets with a 3-D printer is preposterous, religion and science can be reconciled. Religion is true. Science is true. God uses both.

Science has shown vast information about God in the past century, and about how he works. We still have the Bible, but God hasn't added to it in about two thousand years. The implication is that God is giving us information through science now instead of through religion.

The extent of preparation for a planet like earth shows astounding attention to detail. The way to have coal reserves is to have the right organisms on the earth at the right time. The way to prepare oil reserves is to place dinosaurs on the earth for millions of years. The earth's core has to be the right temperature. The crust has to be the right thickness. Everything is in the right place and the right order.

We have an abundance of scientific information because God wanted us to understand how he works. We have the Bible for the same reason.

DNA is a compelling example of God showing us his science. We didn't know in 1,200 B.C., when the Bible was being written, what the function of DNA was. When we mapped the genome, it became clear the shocking level of technology and detail God uses in creation of people and planets. Our greatest technologies are a poor mimicry of his science. We still don't know as much about DNA as he does. Our scientific efforts are pathetic by comparison.

The most astounding miracle of God is that he could use the amount of science he has, including DNA, and still leave us alternative explanations, such as religion and evolution, or a combination of both. He has effectively hidden himself from us using astounding levels of science. With religion, he has hidden himself behind prophets and two thousand years of distance in time.

Science has tried to use data and information to show that everything in the universe can be explained without God.

It is essential to our freedom on this planet that God hides from us. If we had indisputable knowledge of him, we would be influenced by his expectations, and no longer truly free. This level of respect for our freedom says much about him, and us. It affirms that he acts out of love for us by placing us as mortals on a physical planet at the beginning of our billions of years of future life.

God the Scientist used biotechnology to develop all species of physical life. God used science in a laboratory, a planet like ours, to get mortal organisms from amoebas to humans. He used bioengineering. We may be able to get a chicken to grow teeth by altering DNA. We aren't the first ones to attempt these biological changes. The information we have is that God has had billions of years to perfect his science. Of course he works by faith, which I define as science, technology, and a lot of hard work. Add to that love and the atonement of Jesus Christ, which are essential to a theory of everything, but which science and technology can't explain.

There are four types of matter we know of: Type A is spirit matter; Type B is mortal or physical matter made of atoms; Type C is resurrected matter; and Type D is transfigured matter. Religion has told us for two thousand years that God had other types of matter besides mortal.

Within recent decades, science has confirmed that eighty

percent of matter doesn't interact with photons, light, or physical matter. Dark matter is not made of atoms. Science knows very little about dark matter, or Type A matter. Religion knows far more about the scientific characteristics of Type A matter than scientists do.

When you die, you are going to have a death experience. You may see a being of light. He will be non-denominational, because God is a scientist. You can view the experience with or without the prism of religion, but the practical, non-faith reality of the experience will be the same. It is a separation of your physical matter self/body from your Type A matter self. You will remain on earth in a different form of matter. You will be a person with a physical appearance, but not a body made of atoms.

Death is not a travel event. At death, you will remain on earth, whether you believe in God or not. If you are not religious, death is not a religious experience. It is the passage from a physical matter existence to a Type A matter existence, without leaving the earth or going anywhere.

The scientific and practical reality of Type A matter is that you will see what mortals see. You will see the same physical buildings, plants, animals, and mortal people you would see as a mortal. In addition, you will see other Type A matter persons, such as ex-mortals. Billions of ex-mortals remain on this planet.

Heaven and hell are the result of Judgment Day. There is a technological divide causing heaven and hell. Judgment day is programmed with better-than-computer precision to divide based on moral, not religious, parameters.

One thousand years after Judgment Day, everyone goes to the planet he or she chose: planet earth, which will be scientifically altered and upgraded to become utopia or heaven, or the second earth, which will remain in the physical state earth is in now, but which will also be utopia because of the evolution (non-mortal, resurrected/Type C technology) of the people on it. God is the agent of that evolution. Judgment Day, or Revelation Day, will, for many, show for the first time, exactly what part God did or didn't play in past evolution, and why Phase II of evolution involves resurrection, which is a higher technology than the old, methodical form of natural selection for mortal plants, animals, and humans. The physical technology of earth is designed to last billions of years, so long as mortals don't pollute it or destroy it. Technologically, part of the purpose of the burning at Judgment Day is to literally cleanse the earth from pollution, slums, and a few other things.

Near-death experiences are annoying. People are always getting sent back to live again. It's like watching the first ten minutes of the world's best movie. You don't find out some of the most important information. In this book, we aren't getting sent back.

America is the prototype of heaven on earth. God saved the greatest information, ideas, technology, art and science a mortal planet has to offer for the finale.

Nearly all the technology normally available to mortals is already on earth. We are at a zenith.

America was saved for the very last moment of the world's history, a few technology-drenched centuries right before the whole planet goes up in the technological flames of Final Judgment.

This book is speculative by nature. It goes beyond the bounds of religion, science, and human experience.

CHAPTER ONE

THE THEORY OF EVERYTHING

THE THEORY OF EVERYTHING

The theory of everything starts with you, as an individual.

You put everything you know in a frame of reference depending on how it relates to you. You are, in one very real sense, the center of your universe. You are the most important person in the universe, at least to yourself. If you didn't exist, it wouldn't matter to you whether God existed.

Your greatest value as an individual stems from your independent mind, and identity, which I believe are eternal, i.e., not created by God. You are in a mortal body because God took your eternal mind, and put it into your physical body. He attached or assigned your mind to the body you are in.

There is a real possibility, even a likelihood, that as an individual, you would be as important as God, except that he was smarter than you, and that he invented technology, and a lot of other things, first. God did not say, "I am the Lord thy God, and I am more important than you." He said, "I am the Lord thy God...Thou shalt have no other Gods before me."[1] Fundamentally, God recognizes we are as important as he is, as individuals. Implicitly, he recognizes our autonomy, independence, and worth. He respects us. He doesn't assume that he can do what he wants to us, or with us, with impunity.

God is the most important person in the universe because of what he chose to do for the rest of us. After he invented all of technology and science, the question was whether he should share them with you and me, and why. The answer is the meaning of life, and the theory of everything.

The key is love. You've heard that before, but it is still true. God based his relationship to us in love. He didn't base it in authoritarianism. If you were the greatest scientist in the world, or in this case, in the universe, other minds, or people, would be more important to you than mere galaxies, or inanimate matter. You would still not want to be alone. You wouldn't really want to have a galaxy all to yourself, much less a universe. You would be dependent on others, if only to add to your own happiness and enjoyment. Fortunately for God, other minds, besides his, existed. Your mind existed.

The central reason God is more important than you or me is because he's done so much for you, me, and everyone else. He's taken simple service to a whole new paradigm. He invented technology, and he's letting the rest of us discover it. He's letting us run this planet, with all the technology we have so far. He lets us choose whether to believe in him, or be atheists, and he gives us all the opportunities of being human, either way. He is remarkably magnanimous and respectful to us as individuals.

We've developed some wonderful theories about philosophy, religion, government, and science. That's the nice thing about being left to run things, almost as if God weren't there. A planet full of technology is at our fingertips. No belief required.

The question is, what will we become? If I am correct, God is a scientist who invented all technology in the past fourteen billion years, and perfected it. Then he shared it with us, as a matter of love and service.

Here are two options, along a continuum: We can become like God, this scientist who lets us all use his technology.

Or we can become mass murderers. Those are the extremes. We have God, who is perfectly good (assuming you can solve the problem of evil), and mass murderers, who do the most evil thing most of us can imagine. God is the ultimate giver. Mass murderers are the ultimate takers. Not only are mass murderers not giving to others, they are actively and intentionally hurting them, in the worst way possible.

I have been fascinated for some time with how we, as a society, can stop mass murder. It is relevant to this book, and the theory of everything, because it defines the boundaries. There is something worse than a mass murder, and that is a son of perdition, but for most of us, mass murder is the worst act we can comprehend.

The extremes of God and mass murderers help us to see how all of us are somewhere in the middle, and which way we should be leaning, or moving. We give, or we take. We make others feel better, or worse. We teach, and learn, and share, or we are contentious.

This determines what we are becoming. A mortal is more than a mind. We become fathers and mothers, sons and daughters, carpenters or kings.

Our theory of everything should include science and technology, but it should also include society, and relationships. It should include not just what we are becoming as mortals, using atoms as the technology we benefit from, but also the second phase of evolution. We are evolving to become something better than mortals. We will evolve to live longer, but not as mortals. The technology exists. We will conquer evil and death, and achieve utopia, with or without religion, because God programmed it that way.

Ultimately, if there were no God, what we become in the future, and whatever else happens in the universe, would be entirely up to us. That puts us in the same position, effectively, as God. We have the technology and science, and the society. We exist. We are real. What do we do now?

STEPHEN HAWKING AND PHYSICISTS

A few years ago, one of the greatest scientists of our era, Stephen Hawking, published a book called *The Theory of Everything.* Except God.

Here is another option: "The Theory of Everything, Including God."

God is omnipotent because he perfected science and technology within the past fourteen billion years.

God is omniscient because he stored all of his data in a comprehensive database, with perfected access.

God is hard-working and patient, because he did all this through science, experimentation, and bio-engineering, not by magic or mythology.

We had no idea how much hard work. Until now.

We want to understand everything.

SCIENCE

The theory of everything should include God and religion, but we'll start with science first.

In *The Theory of Everything: The Origin and Fate of the Universe,* Hawking describes the universe in terms of science, particularly physics.[2]

For physics, the beginning refers to the big bang, theoretically the beginning of time and matter. Hawking looks from that beginning, to the present, and predicts the future.

For physics, the theory of everything is limited to physics. It seeks to define a universal theory which can explain everything from quantum mechanics to galaxies, from the Planck scale to the cosmos. It is an attempt to define scientific law.

If you really want a theory of everything, and not just physics, you'll have a lot to add.

EVERYTHING BESIDES PHYSICS

For the theory of everything, we add to physics, with its history and future of the universe, all other scientific data and information: biology, DNA engineering, geology, anthropology, astronomy, etc. We accept the facts, and we contemplate the theories.

Then we add politics, law, government, philosophy, history, etc.

We add the arts: music, painting, sculpture, literature, etc.

We add medicine.

So far, the theory of everything includes everything secular.

To all this science, philosophy, and secular knowledge, we add religion. We add God. We add the Bible. We add all of Christianity, and Judeo-Christian beliefs, and all world religions.

We might wonder whether that is really possible. I think it is, and in fact, it simplifies and clarifies things to place them as part of a theory of everything. It also helps to be flexible in those of our beliefs which don't fit in the big picture.

The good news is that the work of developing a theory of all of these things, in fact, a theory of everything in the mortal world, has already been mostly done for us. We have already defined and accomplished so much in these areas. It feels like we are at a zenith, or a pinnacle. We feel like we know almost everything there is to know, and we might be ready for the end of the world. The mortal world, that is. After the mortal world comes phase two of evolution, on beyond Darwin and his progression of animals and humans made of mere atoms.

In terms of physics, we've discovered we only know about a fifth of what there is to know.

DARK MATTER IS A HUGE DEAL

The discovery of dark matter should shake the foundations of science and philosophy. It is the vindication of religion. It shows us the dividing line, clear and bright, between fairies and angels. Fairies are known to be figments of our imagination. Angels are technological beings. We were given this information through eye witnesses thousands of years ago, but now our understanding of dark matter and science shows us that God could have a complex social order not made of atoms, which constitutes eighty percent of the universe. We only know it as dark matter, because he is hiding eighty percent of what he has done. His social order includes mortal planets and utopian planets. He has told us, in religion, about the utopian planets, but most of us are too stubborn to believe.

Dark matter is a misnomer. It isn't dark. It simply means matter scientists only know exists because of its gravitational pull on matter we see.

The premise of this book is that spirit matter, transfigured matter, and resurrected matter are real matter, and likely make up the eighty percent of the matter in the universe we call dark matter.

Science has shown that there are other types of matter beside matter made of atoms, matter which is very much real, unlike fairies.

We aren't talking about beings in another dimension, made of atoms like we are. Dimensions have nothing to do with it. Dimensions are for mortal types of matter, the quarks and other particles which make up atoms. We are talking about beings who are made of a very real type of matter that doesn't interact with atoms. There are billions of these people, ex-mortals, on this planet, and we do not see them or feel them, and that isn't because they are in heaven or hell, it is because their technology is different from ours.

The implication of dark matter, i.e., that it shows us the potential reality of what religion has told us for thousands of years, transcends science and religion. We could stop thinking of churches as mere religion. Churches have been a tremendous benefit to us, in Western thought, morality, and art, a balm against depression for the masses, or yes, even an opioid, a pain-killer, but in a good way. The pain of life is prominent for all of us. We all suffer, including believers. And atheists are left without the pain-killers, unless they look to human relationships and love as the source of meaning, which is exactly what God does. He isn't just a scientist, he's a philanthropist.

We know a great deal about matter made of atoms and photons.[3] Unfortunately, now we might be stuck. Dark matter still has science baffled. The Large Hadron Collider continues to push the frontier of what we know. Still, science only understands about 20 percent of everything, which makes scientists eighty percent ignorant about our universe. This a great reason to take religion seriously, which described the other eighty percent as much as two thousand years ago, by describing spirit matter and resurrected matter. The other great reason to take religion seriously is that our advanced state of society, approaching zenith, is built on religiously-based Western philosophy and thinking, a peculiarly strong mix of secular and religious thought.

Religion is undeniably powerful. It has shaped our thought and culture. While this is mostly a good thing, there is a down side. Religious persecution is alive and well. This isn't good. Religion also may limit our willingness to accept scientific information and data.

TECHNOLOGY DESCRIBED BY RELIGION

Religion says God had technology to view past, present, and future, thousands of years ago. Thanks to Einstein, and general relativity, we understand more now about why that may be possible.

God placed better-than-digital videos of the future in the minds of mortals he called prophets. That is a type of technology. Those prophets told us about it. The Bible is full of this information. Daniel had an important view of the future, including the history of the worldly kingdoms, with a giant statue, and a stone cut without hands.[4]

Joseph, in Canaan, saw a vision of the future. In a dream, Joseph's father, the great Israel, and Judah, Joseph's older brother, bowed down to Joseph.[5] Joseph was sold into Egypt, and eventually saw the future again. Lucky baker, poor butler.[6] You know the details. Again, Joseph saw seven years of plenty, followed by seven years of famine, which helped save Egypt, and also Joseph's family.[7]

Some people think those prophets were wrong, or that the events never happened. Now we find out that scientists think that time might be relative, and that once we perfect science, we might be able to view past, present and future. Using everything we know, combining science and religion, we gain a glimpse of the theory of everything.

There is additional technology. God had a technology called the Urim and Thummim. He gave the technology to people called prophets, or seers. Aaron, Moses' brother, was a seer, and he had a Urim and Thummim.[8] A seer had a technology better than prayer available for enquiring of God.[9]

Consider the day of Pentecost, and then put it in modern technological terms: You interpret languages using your cell phone, and you use Google translate, or a similar app. God had similar technology on the day of Pentecost. Eye witnesses documented the technology: men speaking, and the listeners hearing it in other languages.[10] The result was high tech, instant translation, because God used something similar to cell phone and language interpretation technology before it was readily available.

ANCIENT SCIENCE VERSUS RELIGION

There was conflict between science and religion even in ancient times.

In all fairness, the Egyptians were good scientific astronomers. Think of the precision of the pyramids. They inspired even the masons who built our capitol building, and the Washington monument.

The Egyptians were great chemists. They were great embalmers, and embalming is a type of technology. It is chemistry. When I was in college, I saw King Tut's exhibit. Or maybe it was his cousin I saw, I can't remember. The chemist who preserved Tut, or his cousin, did a fabulous job.

Pharaoh listened to Joseph in Egypt because Pharaoh was willing to consider all data, and pick out the best, and what Joseph had, a technological vision of the future, turned out to be the best data. God allowed Joseph to use higher tech. We confuse it with, or dismiss it as, mere religion.

Perhaps we skip embalming dead guys, which is relatively low tech, and go straight to transfiguration, which allowed Moses to "die", i.e., get transfigured, and then to show up over a thousand years later on the mount of transfiguration.[11] Or is that not technology? Mere religion? Old fashioned? Or should we call it sci-fi? Perhaps we'd rather be intentionally ignorant. We might be too busy ridiculing to acknowledge the power of millennia-old religious data.

Awesome technology abounds in the Old Testament, and Jews believe all that, even today. Strongly. Devoutly. And for good reason. And so do all the Christians.

One more example: Consider Shadrach, Meshach and Abed-nego as using flame-retardant technology. And not even smelling of smoke.[12] That isn't flame-retardant. That is transfigured. That is high-tech. I think that is what is going to happen at the Second Coming of Christ. Those people who aren't liars, thieves, or murderers will get transfigured, and caught up, technologically, so that they won't burn.

THE TECHNOLOGY OF GOD

God can be defined in terms of his technology, instead of his religion. It is possible to transcend religion and science. Religion and science fuse into one.

If you don't believe in God, this book is a framework to explain everything without religion, in terms not just of technology and science, but also of sociology, law, and politics.

For believers, the theory of everything starts with religion, and should also include science.

Either way, we universally try to find the meaning of life.

THE MEANING OF LIFE

These questions define the meaning of life: How did we become mortals? What are we supposed to do? What happens after we die?

THE THEORY OF EVERYTHING: IDENTITY, RELATIONSHIPS, AND MATTER

The theory of everything starts with identity. Who are you?

It next establishes relationships, and ultimately shows that God exists, and has perfected relationships. By so doing, he has also perfected government and philosophy.

Finally, the theory of everything shows what matter is, and how we can put it to use, perfectly. It also shows why starting with good and evil, peculiar to mortal matter and atoms, is the perfect start.

IDENTITY

In my opinion, you are a mind. You, like God, have always existed. You have an identity. It belongs to you, not God, and he respects that. You existed as a mind before the big bang.

God also existed as a mind. He existed before he started developing or creating matter.

Identity is independent of matter. You are more than a human, more than an object made of atoms.

RELATIONSHIPS

If you were all-powerful, i.e., omnipotent, the most important thing to you, besides your identity, would be your relationships to other minds.

I define a mind as a person. A mortal is a mind, or person, with a mortal body.

Animals are also minds, but they do not have conscience.

Religion tells us that God defines us as his spiritual children.[13] A spirit child results from technology: it involves attaching or assigning a mind to a body of spirit matter, which I will refer to as Type A matter, or Type A technology.

If you don't believe in God, you don't accept on faith a relationship to a being you can't prove exists. Without a god, you have instead a first scientist. If we assume humans are the first scientists in the universe, so we have the burden and blessing of defining our own theory of everything.

PERFECT PHILOSOPHY

God has perfected philosophy. He is the perfect philosopher.

The first rule for God: Don't hurt anyone. Don't hurt any mind, animal, or person. Don't offend or interfere with anyone else's rights or autonomy without his or her permission.

The key to perfecting philosophy is love. First, it is necessary to have all power, i.e., all technology, and second, to have perfect love for all other minds or beings. This allows God to always act in the best interests of those minds or beings. God avoids becoming the supreme dictator because of love. He perfectly respects the autonomy of all other minds. We would hope for nothing less from a scientist who has perfected all technology.

The solution to the problem of evil is that God has a contract with every person, animal, or mind who is placed in a mortal world. That contract says the person or animal agrees to be subject to pain as part of a mortal life. They are also subject to amnesia. Causing someone amnesia would be a highly offensive thing to do, unless he or she agreed to it. God establishes a contractual relationship with them. They obtain a mortal body, but only if they agree to be subjected to suffering.

Satan is a being who established a relationship with God, by becoming an angel to God. He then refused any further contract, and was technologically cast out.

Science and technology do not and cannot explain the atonement of Jesus Christ.

FOUR PARADIGMS OF MATTER

Religion teaches there are four types of matter: Type A is spirit matter; Type B is mortal or physical matter made of atoms; Type C is resurrected matter; and Type D is transfigured matter. Religion has told us for two thousand years that God had other types of matter besides mortal. Within recent decades, science has confirmed that eighty percent of matter doesn't interact with photons, light, or physical matter. Dark matter is not made of atoms. Science knows very little about dark matter. Religion describes far more about the scientific characteristics of Types A, C, and D matter than scientists can.

If eighty percent of matter is not made of atoms or photons, there may be more galaxies made of resurrected matter than there are made of atoms. That is, of course, a huge speculative leap, but a possibility; arguably, it is a probability, if the big bang was fourteen billion years ago. Even after mortal science has achieved a thorough understanding of matter made of atoms, science has reached a new frontier showing the extent of human ignorance, with eighty percent of matter unknown to us.

The good news is that our experience as mortals is highly effective and condensed learning. We may not know what the other three paradigms of matter are like, but we won't necessarily have to relearn all of the information about science, art, history, philosophy, relationships, and governments we already know.

In other words, a mortal world made of atoms is like a theme park, or a youth camp, where you can do just about anything, without billions of years' worth of consequences. Unless people really do get sent to an eternity in hell for becoming liars or thieves, which I don't believe. I do believe in hell, and I feel certain it isn't worth going there. Given that most of us repent without going to hell, there is room for human error, lots of it. We call a resurrected world, i.e., the sea of glass and fire John the Revelator describes, heaven, or utopia.[14] It is a perfected world. Whatever you want to screw up, do it as a mortal, since you won't be able to later.

HEAVEN AND HELL GO DIGITAL

Almost everything about heaven and hell can be described in terms of technology.

At Judgment Day, heaven is a technological paradigm shift. It is a dose of data: Jesus Christ is God, and he did everything through science, not magic. Going to heaven could as easily be described as receiving a technologically upgraded planet, and being in the presence of Jesus Christ.

If we really had a theory of everything, we'd get rid of the stuff that doesn't make any sense.

I don't believe we can accept a god who does things by magic. If God created the earth in seven mortal days, he did it using science. It would be possible to create the earth in seven mortal days, but it is improbable. It would take something like a 3-D digitized planet printer, although it is so far beyond our current science that we struggle to conceptualize it. My opinion is that God probably took about four billion years to create this planet. I'm impressed by his patience, and his love, especially if mortals will only occupy this planet as humans with conscience for six thousand years, as religion says. Four billion years is a lot of investment for six thousand years of mortals. Eighty billion mortals, a speculative guess, must be worth it to God.

There is a big difference between magic and faith. Yes, God creates planets using faith, which I define as science, technology, and a lot of hard work.

Then there is the flood. I believe the scientific data, and I believe the Bible. I have an opinion about how the Bible should be interpreted: we should agree to disagree.

CHURCH DOCTRINES I DON'T AGREE WITH

I don't agree with the church doctrine that you can go to hell for believing in the wrong religion.

I don't agree with the notion that the moment you die, you are eternally damned to hell, or not.

This book is non-denominational. It goes beyond the doctrine of any church.

SPECULATION

I love speculating. I relish it. I savor it, and bask in it.

There is a power and freedom in being able to disregard anything taught by a church in the past 1,900 years. It is a useful technique and tool.

This book is full of speculation. It has to be. I'm speculating there is life after death, but I can't prove it. I'm talking about three paradigms of matter, other than mortal matter made of atoms. I am talking about everything, from big bang to eternity.

I am asking readers to contemplate deeper things, in fact, the deepest things of eternity. What would you do if you were God? What would you do if you were an omnipotent, omniscient, scientist? Would you be a dictator, or a father?

CHAPTER TWO

THE PHILOSOPHY OF EVERYTHING

WHAT WOULD YOU DO, IF YOU WERE GOD?

God is the Great Philosopher. He hides from us so that we can feel what it is like to be in charge.

In addition, we existed as minds independently of God, and he respects our autonomy almost absolutely. The exception is his policing function.

The issue of meaning in our existence is as real to God as it is to any of us, including the great philosophers, from Socrates and Plato to Kant and Friedrich Nietzsche. Without God, we can ponder what our meaning would be, and our greatest philosophers have written extensively about it.

For believers, particularly churches, we tend to define our meaning by what God thinks of us, or what he tells us we are. We are his creations, or his children.

To go a step beyond that, God as a scientist would have discovered, after he figured out how to create suns, planets, galaxies, and humans, that he was left to himself to define his meaning. There was nothing greater than him. He defined his importance as his relationship to us: our father, the father of our spirits.

My view of God's philosophy is: In the beginning, we were minds, independent of God. We accepted a relationship with him and became his spirit children. He also gave us mortal bodies, placing our spirits and our mortal bodies together as one. The power of having mortal bodies, with the accompanying good and evil, was to define morality, and make it relevant to us through experience. Arguably, the existentialist view is compatible with this notion.

The theory of everything should show what you, or any highly intelligent, rational being would do when faced with all the technology in the universe, perfected. It isn't about God, it's about you. Would you give a planet to mortals, and give them amnesia, so they couldn't prove what you were up to? Of course you would.

All of religion fuses with natural scientific laws, i.e., gravity, electromagnetic force, strong and weak forces, joining philosophy, politics, sociology, and law to form a comprehensive view not just of how God did it, but why any scientist would do the same. We would do the same, if it were us, instead of God, in charge. And the reason God hides from us, and he does, to a highly effective degree, is precisely so that we will be in charge, instead of him. It is the ultimate empowerment: "I'm God, but I'm going to hide, so that you can think and act like God. Here's a planet. Do with it what you will. I'll give you religion for guidance, along with a few other important things, but for the most part, you're on your own. You won't even know for sure if I'm here."

The way the mortal earth is set up allows us to be in charge for a window of six thousand years of recorded religious history. It is a story of escalation, from a small initial population with limited technology, to the end of the earth with a zenith of science and technology.

It is a pattern, one which can be repeated on planet after planet. Mortals are given the chance to become God-like by being subjected to good and evil.

The central problem of evil, i.e., how a being like God could subject us to evil, without being guilty of evil, is solved through contract law.

God is the Great Scientist. He is also the Great Lawyer, the Great Philosopher, the Great Sociologist, and the Great Technological Entrepreneur.

That is why this book is the theory of everything.

THE GREAT PHILOSOPHER

I was fascinated by The Myth of Sisyphus, in which Albert Camus, a French philosopher (1913-1960) described absurdism. He concludes that suicide is the wrong answer. He won a Nobel Prize for Literature.

The possibility that God exists, and yet hides himself from us, can be problematic and painful. Atheistic scientists and philosophers are essentially trying to figure out the meaning of existence, defined as mortal life, if God doesn't exist.

For believers, God may be the meaning of life. Whatever he wants must be good for us.

My view is that God is a scientist, philosophically. He wants us to experience what it is like to be in charge of our own destiny, as if there were no God. He also gives us religion, if we choose to believe.

God is the Great Philosopher. He speaks through religion, using prophets as witnesses, prophets in the distant past. He shows us his science. Through it all, he allows us autonomy. We don't even have to believe in him. He used his technology to give us a planet, where we are in charge, and we can, and generally do, ignore him, in our governments and culture. Ultimately, he uses a moral code, human conscience, to provide a perfectly fair result for all, regardless of religious belief. He makes religion irrelevant, except that belief provides comfort.

GOD IS WILLING TO BE YOUR LEADER, SORT OF

God is willing to associate with you, and mentor you.

You've got prayer, the Bible, and the Holy Ghost. If you are a Muslim or a Hindu, you have your own sacred texts and manner of worship.

If you don't believe in him, God also respects your autonomy enough that he is willing to hide himself from you. That is a great gift, the gift of freedom, independence, and autonomy. The gift of seeing what it would be like, not if you had never been born, but if God had never existed.

Nearly every society in the history of the world has been able to worship, or not worship, God, in whatever way makes sense to its people. China, India, or Africa, through thousands of years, chose their own religions. This doesn't mean there has always been religious freedom. Some societies have enforced a mandatory state religion.

The rare exception to a non-discriminating God was Israel, in which God made a covenant with people, and sent prophets and angels to them. That was high tech, and also resulted in high demands.

God sending prophets and angels has been rare in the last two thousand years. In spite of this, we have a strong moral code in America. We have the rule of law. We have conscience, and an awareness as a culture of the need for kindness and service to others.

We are at a zenith of history and society, in every way. This includes transmission and broadcast of all religions around the world. Now, many people one in the world can seek out God, to establish a covenant, as Jews, Christians, or whatever else. Even if you aren't of Jewish bloodline, you could convert to Judaism. You can take your pick of all the religions in the world.

God will give you guidance, in whatever way you want to seek him.

I believe there are some Catholics, Protestants, Hindus,

Muslims, Jews, and people from various other world religions who literally know God exists. Most strongly believe God lives, with or without knowing.

The issue is whether God, with his perfected technology, can make it known to some people that he exists. Religion says this is done through the Holy Ghost, or through the spirit of God. This could be compared to a communications technology, but I prefer to leave it in religious terms. God does have the power and technology/religion to make himself known.

Some people are deceived as to some religious beliefs. This doesn't change the fact that some people know God exists.

At the zenith, God has detached himself from the power of technology and religion to some degree. We need to understand that technology could be ours, and not his. Technology belongs not to the first person or scientist who discovered it, but to whoever is using it now. It is transferrable. It is independent. He is the creator, meaning he discovered and invented all technology, but we are in the same position he is, at the zenith of mortal technology: We have the technology, and the issue is what we'll do with it. We can be responsible with it, and use it in love to bless and help and serve others, or we can use it to harm others, start wars, and become despots. We can use our technology and knowledge for good, or evil.

If you don't want to believe in God, then you are putting the burden on your own shoulders, as if humans in our day were primarily responsible for everything in the universe.

MASS MURDER, THE OPPOSITE OF GOD

To me, one of the most compelling aspects of philosophy is the contrast between God and mass murderers. The knowledge that a man could go to Vegas in 2017 and start shooting people at a concert brings everything into an ultimate focus: meaning derives from kindness, love, and a positive relationship with other people. Love is God's motive, and mass murder is the opposite of that. It ignores the importance of human relationships, and the independence of the human mind. God created us, but a mass murderer can destroy us. A mass murderer who doesn't believe in the afterlife can believe he has the God-like power of destroying us. The theory would be that evolution created us, and some putz of a mass-murderer can obliterate us. This would be nihilism, and it is a singularly unpopular idea because it is so contrary to the information and data we have from life.

Atheistic scientists are not telling us that life is meaningless, or that mass murder doesn't matter. Such a position is indefensible. It is nonsensical. The most famous skeptics of religion don't go around encouraging suicide or mass murder.

Conscience matters, as a universal code of conduct, and it is undeniable. Morals matter, to religious believers, and to atheists and skeptics of religion. Being nice matters. Being kind matters.

Many people find meaning in religion. The point of religion is to worship, or become more like, God, and to obey him.

Unfortunately, religion can be a motive for mass murder. It is possible to believe that God will reward you for mass murder, if you are doing his will. That is a powerful and dangerous notion. The primary danger is that God has nothing to do with it.

WE HAVE VALUE INDEPENDENTLY FROM GOD

Some believe God creates us, and gives us our meaning.

I think the notion that God gives us our meaning is an inferior option.

The better option is that we have meaning and value

independently from God. We are empowered by the notion that we are independent of God, at least our minds are. We have an identity independent of God. That is why the soul feels its worth. It isn't because God loves you, although he does. He doesn't have to validate you. You validate yourself. You are a mind, independent, with personality and identity. God is described religiously as "I am," the self-existent one. We also are self-existent, which gives us tremendous value, a value recognized by God. That is why he goes to all the trouble to give us high-tech mortal bodies, and a nice planet to live on. We are worth it to him. God defines his own meaning using love and relationships. He is a father, not the great dictator. In fact, we give God meaning. Without our minds, what he would be? Alone, and meaningless. Even if he could organize all the galaxies without help, he would be left alone. He might get as depressed as Friedrich Nietzsche.

THE MOTIVE FOR MASS MURDER

It is shocking that anyone would commit mass murder. The Las Vegas shooter was an example of how shocking. He planned very well to kill a large number of people. He wasn't crazy. He was intelligent. He was capable. He was effective. He had a well-planned strategy.

We are disturbed that a mass murderer could become so convinced of a false idea, that he would kill people, and that he could think that the more he killed, the better. The Las Vegas shooter had a goal to kill as many people as possible, judging by the number of weapons, location, and strategy he chose. He appears to have planned to set a record of how many people he could murder, at least as a single shooter.

There is a motive for mass murder. It isn't as simple as greed or need, the motives for theft. It includes need, the need for recognition, for importance.

One of the most powerful needs, as demonstrated by mass murder, is the need for power, control, meaning, attention, recognition, and self-worth.

A mass murder is motivated, in part, by the idea that the murderer will achieve meaning through the greatest violence and harm he is capable of. He will become important. That notion is dangerously compelling, which surprises most people.

Part of the powerful, and dangerous idea (dangerous enough that more than just a couple of mass murders have been committed) is that life is meaningless, so killing people doesn't matter. Their pain doesn't matter. Their families' pain doesn't matter. The notion is related to nihilism.

The motive may go beyond just the belief that it doesn't matter if you kill someone. It may include the notion that people are bad, and they deserve to die.

Mass murder has additional motives. One is anger, and the idea that some people deserve to die, or that all people do. It may be an act of punishment against people who have injured or spurned the mass murderer. It may also be an attempt to fight back against existential pain, the daily pain of seeing no meaning in life. Presumably, most mass murderers aren't angry at God; they don't believe in God. They are angry at the constant, absurd pain of mortal existence, the apparent meaningless of it. They are also making a point to God, if he exists, that they are daring enough to defy him. The tragedy of this is that God is perfectly loving, and defying him has no meaning. Hurting others is the meaning, and the worst possible trap of misery.

Many philosophers have described the pain of life, at length, including Friedrich Nietzsche. Albert Camus, the French philosopher, wrote of the absurdity of life, but fought against nihilism.

From a dark standpoint, God causes us a lot of pain, even if we are believers. There are some really good reasons to be angry at him; it takes a lot of faith to answer them all. Most of us believe in his goodness so strongly, we couldn't get mad at him if we wanted to. Like Job, we would praise him while we suffered, even if he killed us.

We also suffer a lot of pain even if we don't believe in God. Life is ugly at times. Life is full of pain. The question is whether it is worth it.

Depression can be defined as being medically unable to be happy most of the time. Realism and maturity can be defined as understanding that life is full of pain, but that the good things in life far outweigh the pain, on a daily basis. We ask the question: Is life worth it? Suicide is caused by answering the question wrong.

HOW TO STOP MASS MURDER

The way to stop mass murder is to convince the potential future mass murderers why they should not succumb to the deceptions and falsehoods that entrapped past mass murderers. No one is harmed as much by mass murder as the perpetrator himself, although no one feels particularly sorry for him, since he is the one causing the damage.

There is a similarity between us, as individuals in a society at zenith, and God. We have a lot of power and knowledge, and so does God.

We have to decide what to with our power and knowledge.

God figured out that having the power to create galaxies was not necessarily fulfilling, if he had no friends or family. He figured out love is the key. Love is the key because relationships with other minds and people are the greatest source of satisfaction and value available, aside from our own self-worth. We are important, because we exist. We are also thrilled to have others around us. They are important to us.

A mass murderer finds himself in a position like God. If believers look to God for their meaning, they should also understand the limitation: God has meaning because he has technology, and a lot of great relationships. Without these things, God would have no more meaning in his life than the mass murderer.

Nihilism, and existential loathing, are absurd.

Anyone who desires to become a despot, a Hitler, will eventually learn that it is a false promise. There is pleasure in controlling the lives of others, and having wealth and power. It is far less, however, than the pleasure that comes from a win-win manner of living. God has sufficient technology to have plenty for everyone. Scarcity need not exist, except on mortal planets, where it is vital to the experience, just as other types of evil are.

It is challenges us, as mortals, to be stuck in the midst of evil, which seems so unnecessary at times, that we may come to believe it is the norm. This is somewhat rational, but incorrect. The balance of good in life is virtually always tipped in favor of being worth it. The exception is during torture, but the better solution, if available, is to halt the torture, not to die.

The mass murderer gets so caught up in his own emptiness, pain, and negativity as to lose sight of the joy that comes through lifting, serving, giving, and helping others.

The lesson for us all is the precious nature of life, families, and relationships.

They are as important to us as they are to God, and for the same reasons. For atheists, the conclusion and mandate should be the same, with or without God: Be kind, enjoy every day of life, serve, lift, and help others, regardless of whether there is life after death.

The tragedy of a mass murderer is that he reacts to knowledge, power, and life with the ultimate act of destruction and hatred, harming others in the worst way he knows how, instead of being like God, who is perfectly loving and kind. God became the best he could be, and remains perfected, fourteen billion years into his existential angst. A mass murderer blows a gasket within the first eighty years of being exposed to good and evil, and ends up becoming the opposite of the most he could be. Even then, there is hope of deliverance, but not until after a thousand years or more of suffering.

SERIAL KILLERS

It is harrowing to read or listen to the words or confessions of serial rapists and killers. Their motives tend to be more clear than other types of mass murder. A serial rapist and murderer is motivated by sexual desire, control, power, revenge, and hatred, among other things.

A mass murderer similarly desires to exercise power and control over the lives of others, who may be strangers or acquaintances.

WAR AS A TYPE OF MASS MURDER

Mass murder can also be caused through a religious belief, or as an act of war. Osama bin Laden attacked Americans on 9/11. We understand it as an act of war, motivated by hatred. We understand the motives of war better than mass shootings. We understand despots who are seeking power, dominance, and wealth. We understand ethnic, racial, and cultural hatred. We understand religious motivation for war, even though we may diametrically oppose it.

SONS OF PERDITION

Religiously, there is one thing worse, arguably, than a mass murderer: a son of perdition.

Sons of perdition are what happens when a mortal gets just as evil as Satan. It has everything to do with hating God. I don't mean atheistic scientists. I mean people who strongly believe God exists, or they know he does, and they hate him anyway. And what's more, they defy him, and would try to kill him if they could.

I understand what it means to hate God. There was a time when I felt enough pain and grief that I hated God. I felt a certainty that he was causing me suffering on purpose. What saved me is that I reacted by obeying God, in spite of my hate. I also never stopped loving God, even when I hated him. It is possible to do both.

SATAN, AND HIS FOLLOWERS

We wonder whether it is possible for someone other than mortals to be truly rotten. What would happen if a technological non-mortal being decided to get really nasty? What would we call him? Satan.

If God forgave Satan, even now, and allowed him to have more technology, he'd just start another war. That's how he is. Evil can become incurable.

MASS MURDER, WAR, AND OTHER EXTREME EVIL

God ends up having to cover for all contingencies of how minds, persons, and mortals, might react to what he gives them. He dealt with Satan, and continues to do so, using technological paradigms to control and constrain Satan. The essential message implied in the technology he has provided Satan with is this: "Go ahead and tempt mortals, but you can't physically harm them. You can't kill them. You can only persuade them to harm each other."

THE GREAT MILITARY LEADER, AND LAW ENFORCER

God is the great military leader. God protects the galaxy, in technological terms. He protects our solar system. If you don't believe in God, feel free to take that for granted.

If you do believe in God, be aware that he doesn't share his power with idiotic, evil despots, like Satan. He protects us from Satan, using technology. Satan is made of a type of matter, Type A or spirit matter, which doesn't interact with humans. Satan has never killed any human, and that's why.

God is also the great law enforcer.

Mass murder happens to mortals. Somebody better have a plan to punish mass murderers, particularly if they try to escape punishment by killing themselves.

Most of us do things wrong, and we don't receive harsh punishment. Our conscience is a technological reprimand, highly effective on a mind, even when it is encased in a mortal body.

All of us sin, and all of us benefit from repentance.

Even atheistic scientists get the benefit of repentance and forgiveness, and not because they pray and ask God to forgive them. It is because they feel guilty, and they change their behavior. For example, an atheistic scientist commits adultery. He feels terrible. He tells his wife. He stops the affair. He doesn't divorce, and over time, his relationship heals. He still regrets it, but he no longer feels the guilt he once felt. I don't think that could happen without the atonement of Christ. I also think that the atonement applies to all, including atheists.

Repentance is a part of humanism, acknowledged or not.

God is in the business of enforcing morality on humans. He describes it in religious terms as remorse and hell, but in a limited sense it is technological. It is like a computer program. By that, I mean it will function just as well even if you don't believe in God. You can still go to hell. The part that isn't technological is the atonement, and you don't have to believe to get the benefits of it. I am convinced that all atheists have benefited from the atonement, even without believing. That is gracious of God.

God does punish us for sins, if we don't repent. He does act as the Great Law Enforcer.

Murderers and thieves don't get away with it. They feel remorse while they are mortal; if they don't change before they die, they continue to feel remorse after they die. At Judgment Day, they are technologically punished. First, they are cast into hell, with Satan and his followers. That is, they no longer get to stay on a nice planet earth. Also, they don't get to repent, and so their guilt continues for a thousand years after judgment.

This police system is technologically adequate, with or without belief in God.

TEN SECONDS AFTER YOU DIE, FOR MASS MURDERERS

I feel a reasonable degree of certainty that when a mass murderer kills a dozen people, and then immediately kills himself, that he will remain on this earth ten seconds after he dies. He won't go straight to hell, for reasons explained later in this book. He will be in a Type A matter (spirit matter) form. We will assume he looks like a person in a white robe, i.e., not naked, and not wearing the mortal clothes he had when he died. He will be surrounded by the dozen people he just killed minutes before, who will also be Type A matter people, i.e., ex-mortals.

The people he just murdered aren't going to beat the crap out of him, though he probably deserves it. Type A matter beings aren't capable of physical violence. They, and any of their family members (grandparents, etc. who died before them), may be angry, hurt, and indignant. They may be filled with shock and hatred, or deep sadness.

I believe the ex-mortals will remain and observe the mortals' reaction, and the aftermath, of the mass murder.

The technology of Type A matter beings includes feeling emotions. I believe this will include a heightened ability to feel remorse, and an enhanced perception and memory. I would be surprised if it doesn't include the ability to feel the pain, hurt, and anger of others. All of this is like digitized information, with enhanced and upgraded technology, not dulled by mortal numbness or nightly sleep.

The mass murderer can sit there and face his victims and their families, or run away, and seek anonymity. Any ex-mortal who loved him, or was acquainted with him, will likely be aware of the mass murder he committed. If his mother died before him, he'll face her, or avoid her, but the ugly truth will remain.

He just has to deal with it.

When Judgment Day comes, he's likely to run and hide. He will be cast into hell for a thousand years, which isn't precisely a lake of lava. Lava doesn't burn an ex-mortal. It is a remorse of conscience, combined with being cast out of planet Earth.

UNIVERSAL CONSCIENCE

Conscience is universal. The only mortals who don't feel any conscience are those who become so hardened, they are numb to it. Sociopaths are perhaps an exception. They show us that it is possible for a mortal body to be ill, to the point of not feeling normal guilt. They still comply, generally, to societal norms, because of the consequences if they don't.

GOD AS THE GREAT PSYCHOLOGIST

If we existed, eternally, as minds, then our mortal state is, in a sense, unnatural to us.

We long for utopia. We delight in symmetry, and think of it as beautiful. Our fashion models are those who fit our views of symmetry and proportion.

Most of us fear public speaking, at least initially.

We have a sense of the passage of time. We feel time as alien, as foreign to us, as grating on our soul, and we wish at times that we could either control it, or get rid of it. We feel an unbearable lightness of being, an existential angst. There was a 1988 movie named *The Unbearable Lightness of Being*, based on a book. I can't say what the book or the movie meant, but the phrase itself had meaning for me. The unbearable lightness of being is what I feel, particularly on Sunday nights, when I long for meaning in my life, for all my problems to be solved, and to know exactly what I am supposed to be doing with my existence. Life's experience has taught me the futility of such longing, but I still feel it.

Religion tells us that we will live forever, if we are obedient to God, in heaven. We define heaven as having no illness or death, and no evil. There is no ugliness or decay, only eternal beauty for persons, buildings, and planets.

If our minds are eternal, and that is what they are accustomed to, then the eighty years, on average, we spend as mortals is unnatural and jarring. Every day we are surrounded by not just beauty, but by danger, death, and decay. We smell our sewers. We see junky old cars, or crumbling old buildings. We don't see the perfection we expect. Possibly the reason we expect, or long for, perfection, has nothing to do with our human, or evolutionary nature, and everything to do with our nature as eternal minds, unaccustomed to the evils of decay and death.

We go to our psychologists and our psychiatrists, and they tell us we shouldn't be depressed. Our universities, as an accumulation of centuries of learning and thought about psychology, don't explain everything. If human identity is actually an eternal mind attached or assigned to a mortal body, our psychology can't possibly explain everything, without looking to spiritual, other-than-mortal, sources.

I believe strongly that I am in fact an independent mind, attached to a mortal body for about eighty years, but in spite of this, I am utterly incapable of ignoring my daily human emotions. I get discouraged sometimes on Monday morning, before I go to work. Once I start working, I overcome the discouragement, and I get a lot done.

Everyone I know longs for attention. It is obvious when your children say, "Look, Dad, watch me go down the slide!" We may think, "I've watched you the first five times, why do I have to watch you again?" Still, we know why they long for our attention. News anchors on major networks, and presidents and senators, long for attention in much the same way. When we are alone and lonely in a nursing home, we'll still be longing for attention. It is a life-long pain in the neck to be that needy of attention and ego-stroking, but that's how it is. I think that is the nature of our mind, which is eternal, and not a function of our mortal biology. The issue is how much of our nature is caused by our eternal mind being stuck in a relatively weird mortal body/brain, which has all kinds of things that can go wrong, i.e., depression, bipolar, psychosis, schizophrenia, etc.

Our view of God as the Great Psychologist, then, might be that he respects our autonomy; he listens to us (prayer) whenever and always; he encourages and pushes us to live, love, serve, and try new things; he is perfectly and endlessly attentive to us; and he wants us to not be discouraged, because things really aren't that bad, ever, from an eternal perspective.

That is what we get from a being who is omniscient, and omnipotent, because he's spent fourteen billion years perfecting science, and learning how best to interact with mortals on a planet. To atheists, he is hiding; to believers, he's ever-present. Could we do any better if we were the ones who perfected science?

GOD AS THE GREAT LAWYER

Add to the God as the Great Philosopher, the notion that God is the Great Lawyer. This answers the problem of evil. We contracted with God, before becoming mortal, that we would have amnesia, and that he could hide himself from us. We signed on the dotted line that our children could die of cancer, or get shot by mass murderers, or that we could get divorced, or run over by trains, or however else we might suffer and die.

To me, this is the only possible rational explanation.

I believe strongly in God, but there are times I have hated him. Like Abraham, I've been tested to the utmost of my capacity. I feel God has tortured me in ways. He has caused me greater suffering than any human is capable of.

On one occasion, during the worst grief of my life, I demanded that God stop hurting me. I rejected his will, and demanded that he change. If I had, at some point before my birth, signed a contract with him that he could inflict so much pain upon me, without getting in trouble for it, I told him that I was revoking that contract, and withdrawing my consent. I told him I intended to hold him responsible.

My effort failed. I ultimately decided that irrevocability was a crucial part of the contract, and that I knew more about it then, than I do now, because of the contract of amnesia.

God cannot inflict pain on other beings or animals without their consent, without committing sin. It violates his principles. Inflicting pain on another, without their consent, is evil. The only way he can get around it is by using a contract with us. He also has a contract with zebras. They get to live enjoyable lives, but they consent to getting eaten by lions.

GENDER

I am a feminist. I want to empower women. For my wife, it was enough for me to stay out of her way. She was already in law school when I met her. For our daughters and granddaughters, speaking of society collectively, we want to see them go to college, get the utmost education available, have great careers in all the many fields, and lead corporations and nations.

We place gender in context as part of the theory of everything, without reviewing thousands of years of culture.

We are at the zenith of culture in the United States of America. Women have rightfully claimed all opportunities that men have, but our society by no means has perfected equality.

We haven't changed what science calls evolutionary characteristics. We have understood that our physical characteristics don't have to define our gender. Our identity, if it is as an eternal mind, arguably could transcend our mortal gender.

The ultimate value of women, and men, is in their eternal identity. It is in each person's mind.

It is remarkable that the Bible placed priesthood, and government generally, in the hands of men. I don't mean it is necessarily good, just remarkable. For thousands of years, there have been reasons and opportunities for societies and cultures to repress women.

At the zenith, we see the possibility of transcending gender issues; of equality among people. We could choose, if we were God, to transcend gender. By the theory of everything, I don't just mean my ideas of what the theory of everything should be, and which I've written; it includes the theory of what is possible in the universe, given the big picture, and given four paradigms of matter and existence, of which mortality is merely one.

Even with all power, all technology, all love, and all knowledge, God still defines himself using gender, which in turn relates to the evolutionary development of humans and animals. Fatherhood became defined by reproduction, as did motherhood, and those were developed, as we understand them, as part of mortal species. They were the essential and chosen forms of reproduction.

GOD AS THE GREAT SOCIOLIGIST, AND THE GREAT POLITICIAN

God, in addition to becoming omnipotent by perfecting technology, had to decide how to control power and politics. He decided to define it using love, with love as the key to everything. He defined himself as our father, instead of as our supreme ruler. He told us his rules, using religion, so we could take them or leave them, religiously. But conscience is technological. It transcends mortality. It can't be ignored. Try and do something rotten. Be violent to a baby. You have to be really hardened to slap a baby and not feel guilty. It isn't because you are exempt, it is because you've learned to ignore conscience. If you do, you will technologically and politically go to hell, eventually, because no organized society can allow slapping of babies without a penalty.

God doesn't send people to hell for not believing in him. He just warns you of how things are, will be, and have to be, in any society, like his, which has transcended mortality. His society has evolved to a utopian state, i.e., a technologically perfected state. Welcome to Evolution, Part II.

EVOLUTION, PART II: GOD AS THE GREAT SCIENTIST

Charles Darwin gets credit for Evolution, Part I. He discovered natural selection, which is an awesome scientific tool. He's been bashed by religion for years, but he can easily be viewed in a different light: The theory of evolution doesn't show why God doesn't need to exist. Rather, the scientific data and information were given to us by God so that we would understand the scientific process God used to develop mortal species and technology.

Darwin was given data, a scientific revelation. He received that information because God wanted him to have it, and for us to have it. God was showing us that God is a scientist, and that he technologically developed living mortal things. God works scientifically, not mythology or magically.

We've got a scientific tool even cooler than natural selection: we have DNA engineering. We have cloning. We've figured out that if you were smart enough, you could bio-engineer everything, starting with amoebas, and ending with man.

God did it. That is what evolution really means: the bioengineering of God. The subtle hint: God looks like a man, not an alien.

How do we know? Look at God's son, Jesus Christ. You can believe in him or not, but if you believe in him, you have to believe God had a son who looked like a human. God uses technology, and after much development, natural selection, bioengineering, and DNA engineering, God ultimately decided a human body is the ultimate form mortal technology can achieve. God had his son, and the rest of us, look like humans. The mortal, human form, however, isn't perfect. You still get hangnails and hemorrhoids.

The most remarkable part, perhaps, is that Jesus Christ,

as God, is going to look like a resurrected human for the next fourteen billion years. He chose that form, a human form, which was originally developed through scientific processes. The implication is that everything is scientific. Everything is technology. God is omnipotent because he perfected science and technology. Resurrection is an upgraded technology, which allows you to life forever, instead of dying in eighty years. Resurrected technology will allow us to look, and eat, like humans, but without getting sick, or aging.

Resurrection is the technology humans evolve to, once they get done with eighty years of the thrills and chills of being mortal. Evolving isn't really the right word for resurrection, since it is instantaneous.

Thank you, Charles Darwin, for your contribution, but you've just been transcended, passed up by Evolution II: resurrected beings. Resurrection is a technology, perfected, and used by God, because it is eternal.

No, you don't want to be eternal as a mortal, no matter how much you love Darwin, science, or (mortal) evolutionary theory. You don't necessarily want to live almost a thousand years, even if Adam did.[15] Be happy with the approximately eighty years you get. Even if we could eliminate all disease and aging, we'd still have problems (at least on a mortal planet). Mortality is a comparatively crappy way to live, eternally.

EVERYTHING IS DIGITAL

God might have created this planet using a 3-D printer, which means the planet would have to be digitized. That's partly what I mean by the wildly speculative concept of digital planets.

There is another meaning: Everything on this planet is digitized. This is a digital planet.

The myriad of colors we see every day are produced with

computer precision by the frequency of the photons bouncing off colored surfaces.

Photons are processed by our eyeballs, and turned into electrical impulses which convey the data to our brain, so that we can appreciate the blue of the sky, and the emerald in the grass.

We can jump off a cliff, and the way we splat is instantaneously determined by physics, with greater-than-computer precision.

Everything is digital, but it looks as low-tech to us as Adam and Eve.

Human beings are extremely high-tech. For some reason we think that things that are biological and alive are "natural." They aren't. They are digitized. Data flows, and the tree grows. Data flows, and ducks fly south for the winter.

DNA is everywhere, from potatoes to whales. We've figures out how highly precise it is, better than digital. It isn't just more complex than our supercomputers can fully assimilate, it also is biologically active. It is like an executable program. Cut your finger, and in a few weeks, watch it digitally repair, the DNA program executing.

As mortals, we are like avatars. We are minds, with the ability to operate and control our mortal bodies. Our bodies function on light and electricity. Our neurons fire, and the messages and data spread through our nervous systems.

At the level of atoms, we are made of a technology of electrons whirling about a nucleus, strongly held in orbit by technology, a natural law, so-called, but which is better described as being like computer programming, only upgraded. We are digital beings, from the atomic level to our nervous systems and brain functions. We are an integration of mind, spirit technology, and mortal technology, in ways we only glimpse, mainly because we can't scientifically characterize the nature of dark matter, i.e., spirit or Type A matter. We are programmed mortally with DNA, and chemicals and proteins perform functions like healing a cut, using technology we can't duplicate, without conscious effort. We take it for granted as biological, instead of seeing it as bio-technology, God-like, beyond our comprehension.

You might believe plate tectonics and earthquakes are determined by nature. I say they are digital. They are programmed, with a technology better than computers. The earth is a digital program, and it is executing, continuously. It is like the automatic thermostat at your house, which changes temperatures because it was programmed. We can mess it up with global warming, but we don't control the program.

THE POWERS OF GOD

God has the power to attach or assign peoples' minds to their spirit matter bodies, and their physical bodies, so they are human or mortal. He does the same thing with animals, by attaching or assigning the mind of the animal to the physical body of the animal.

At death, the mind stays attached, and wired into, the spirit matter/Type A matter body.

God also has the power to attach a mind to a resurrected body, permanently.

This is one of God's greatest technologies: to attach or assign minds to physical bodies, to create living mortal things. Our current science has no knowledge of how life began.

The powers of God include technologically producing spirit, mortal, resurrected, and transfigured bodies, and attaching or assigning minds to those bodies.

The birth of a mortal child is a well-masked technology. We tend to view it either as God's miracle (magic), or as a biological process. By contrast, the technological nature of resurrection is more obvious. Religion describes that Jesus Christ was perfectly restored, with a human appearance, after his death. His apostles and followers saw him, and talked with him.

The ability of God to perfectly repair a decaying body, in the case of Lazarus, shows the digitized nature of the information, i.e. DNA, reconstituted in a mortal.[16] It shows remote restoration and reconstruction, like a 3-D printer, pervasively through every cell of a human body.

GOD LIMITS US, AND WITHHOLDS INFORMATION FROM US, INTENTIONALLY.

If we could exist as minds, going back billions of years, could we reverse-engineer what God did, knowing what we know now? We don't have a way to effectively investigate what it would be like to have a mind with no physical body, because God stuck us in a physical body. According to religion, after we die, and before we are resurrected, we will exist as spirits. In all likelihood, being a spirit has a different set of limitations, placed by God.

God was a mind, in the beginning, before he was a spirit, and before he was a mortal, if he ever was, and before he became a resurrected being, if that is what he is now. He may have had the ability to interact with all types of matter in ways that he has barred us from doing, by placing us in mortal or spirit bodies.

As a mind, God may have been able to interact with matter at all scales and levels. He may have been able to view atoms and quarks directly, and to engineer them in ways that are incomprehensible to us, technologically, because quarks are so small compared to us. His technology may include assigning minds to matter at the smallest of scales, to implement technologically gravity, electromagnetic, strong, and weak forces. We see how he works with humans: he assigns or attaches minds to physical bodies. If he has the same ability at the neutrino level, and has everything digitized, with better-than-computer precision, we can glimpse how he developed and perfected all technology and natural law.

God gave us all amnesia, which is a rotten thing to do, unless we agreed to it before we were born. I value knowledge. If I had some really important knowledge, like whether I existed before I was born, and someone took it away, I'd be angry at them, even it was God. He'd need a good reason, and he has one.

When I was in high school, I heard that the U.S. military used to put a governor in the motors of some jeeps, so they couldn't go faster than a certain speed, like 55 miles per hour. That's what it is like. God has placed restrictions on us. He controls how much we can get away with by placing us in mortal bodies. He does it using technological paradigms. These are just as effective on atheists as they are on believers.

We are tightly controlled by God, in a way. Perhaps it is more accurate to say he limits what he gives us. By giving us mortal bodies, he gave us a lot. He gives us the joy of being fathers and mothers, which in my view is unbeatable. He also gives us the chance of becoming serial killers, or murderous dictators of countries. In spite of giving us so much, he still has strong limits on some knowledge.

I don't agree with the doctrine of some churches that we are 100 percent created by God. It devalues us. We are minds, which were independent of God, and one of his greatest defining virtues is that he loves us enough to give us almost boundless opportunity, while refusing to compromise our autonomy.

GOD LIMITS THE TECHNOLOGY HE GIVES TO IGNORAMUSES, TO PROTECT US

If we were minds, floating in darkness prior to the big bang, and we were as smart as God, we would have been figuring out how to create matter in an orderly way, i.e., the big bang.

What is to keep us from doing that now?

Our physical bodies. We are very much stymied.

After we die, will our chances of sneaking technology from God improve? No, he carefully guards it at all levels. Consider Satan.

Satan doesn't have a physical body. He is a technological being made of spirt matter, very real, but he can't interact with mortals, except by putting ideas in our minds. Temptation is technological.

Satan hasn't figured out a way to beat God's technology. Satan is trapped on this planet. The reasons are technological.

CHAPTER THREE

THE TECHNOLOGY OF GOD

There has been a stunning convergence of religion and science in recent decades, for those who believe strongly in both. In this chapter, we will look at the issues in greater depth.

We long for the day when our scientists and doctors can cure cancer and prevent aging. Our utopian dreams of the future include having adequate crops on the earth to eliminate hunger; adequate housing for all; total elimination of all disease; and lengthening our life spans into hundreds of years, defeating aging. We've made great strides in science in the past several centuries. We marvel at what we might accomplish in a thousand years, or a million.

The combination of religion and science now shows us that God has already had fourteen billion years to do all that. He has already perfected all science. Our greatest technologies are a mere mimicry of his. Our supercomputers guide missiles, and rely on atoms, molecules, and electricity to store and transfer data. His supercomputers order the galaxies. God's works are eighty percent hidden from our science. He is not just a scientist, he is the Great Scientist: The reason God is omnipotent and omniscient is precisely because he has perfected all science.

God orchestrated the big bang fourteen billion years ago. He existed before, during, and after the big bang as a mind, which is not made of physical matter, and was unaffected by the intense heat of the big bang. He gave us a recording of the data of how he did it, and of everything he's done since. The recording includes the microwave data. It also includes light from trillions of stars, which travels to us at the speed limit God placed on photons so the data would reach us in a way we could understand it. He hasn't shown us how he introduced gravitational fields and electromagnetic fields, but we know the calculations required for the formation of physical matter during the big bang were astoundingly precise. We don't know why the strong force and the weak force organize atoms, we just know that they do. Scientifically, we don't know how life exists, we just know that it does. Religiously, we know that God puts a spirit in man's physical body, which makes him alive. Putting life into mortals is a type of knowledge. God knows how, and we don't.

Our earth is a vast recording device. It recorded the data of the creation, four billion years' worth.

Our mortal bodies are astounding recording devices. The information found in our DNA and cells is all recorded and available for study.

God is defined by the way he works. The vast recorded data he has provided shows us the level of his science. We have improved science in the past millennium. In fourteen billion years, God has perfected science, and we see it in everything from our galaxies to our quarks.

THERE ARE FOUR PARADIGMS OF SCIENTIFIC MATTER

The Bible is full of information about the technology and science of God. The obvious example is that the Bible speaks of spirits and resurrected beings as having different characteristics from mortals. In recent decades science has confirmed that eighty percent of the universe is composed of dark matter which is not mortal physical matter made up of atoms, protons, and photons.

Religion has taught for over two thousand years that there are four types of matter: spirit, physical, resurrected, and transfigured matter. They are "real" and scientific; there is no such thing as immaterial matter. The inability of our science to ascertain the characteristics of three of these types of matter stems from our scientific ineptitude, not from lack of reality. Type A matter is referred to in the Bible as spirit matter. Its characteristics include that mortals can't see it or feel it. A being made of Type A matter can be among mortals unseen and unheard by mortals, but also able to see and hear mortals.

Type B matter is the mortal, physical matter we are familiar with. It is made of atoms. It includes physical light, or photons. Science has shown us an astonishing amount of information about this matter, from quarks to galaxies.

Type C matter is referred to in the Bible as resurrected matter. The Bible speaks of Jesus Christ as a resurrected being: he was able to talk with his followers. They saw him and felt his hands. He ate mortal food before them, fish and honeycomb. He entered the room without opening the door.[17]

Type D matter is primarily reserved for Judgment Day: Religion says that mortals who are not going to hell will be transfigured, or temporarily protected by a physical change, so that they are not destroyed by the heat.

It is stunning to think that religion described scientific information and characteristics about three other types of matter two thousand years before modern science discovered dark matter.

Light is not limited to photons. There is a light associated with appearances of beings made of Type A, Type C, or Type D matter, which is light with scientifically different characteristics than photons.

Communication is far more efficient when it is not made using mortal physical means. Light has a speed limit, scientifically proved for photons. Other types of light do not have the same speed limit, based on what religion says. The Bible calls instantaneous communication with God prayer. Obviously, it is not limited to the speed of light. Science might argue prayer can't be real because of the speed limit on physical matter and photons. Religion describes a much more effective means of communication, in fact a scientifically perfected means, not based in physical matter communications like sound waves, radio waves, electricity, or photons.

God is defined for Jews in the Torah, and for Christians in the Bible. God has given the nations and cultures of the earth wise leaders who have established world religions, and made people better than they would otherwise be.

Religion says we start on this planet for eighty years, and then spend the next fourteen billion plus, i.e., eternity, as non-mortals.

HEAVEN AND HELL ARE SCIENTIFIC

Heaven and hell are scientific. God gave us parameters in religion in the Bible. There is a list of people who go to hell. The hell list includes those who ignore the universal moral code known as conscience and become liars, thieves, adulterers, and murders. The judgment of God about who goes to hell is made with better-than-digital precision. The hell program is so pervasive it is infallible, and will result, scientifically and morally, in the placement of those persons in hell who deserve to be there.

Near-death experiences provide us information that after death we look like people, but we are composed of Type A matter. The overarching experience of near-death experiences in cultures across the world, if there is a being of light involved, is that the being of light is non-denominational and non-judgmental, which is consistent with religion.

Religion, which I will define as the Bible, says when we die we go to the spirit world (using words like sheol, hades, paradise, or hell). The spirit world is on earth. Satan and his followers are also on this earth.[18]

Digital heaven and hell means that everyone who dies, perhaps as many eighty billion people, and forty billion of Satan's followers, are scientifically Type A beings inhabiting this planet, not interacting with mortals any more than dark matter does, except that temptation is a scientific form of communication between Type A matter beings and mortals.

At Judgment Day, religion says God will appear to the whole earth all at the same time. He is scientifically going to produce so much heat that the earth is to melt. Any physical matter will vaporize at five thousand degrees, so the physical temperature will be lower than that. Otherwise the earth would turn to gas instead of melting. People vaporize at less than five thousand degrees. Any mortal on the hell list will vaporize at that time, and join the Type A matter group with other people who died on the hell list. These people all remained on earth when they died.

The Type A matter group are not going to vaporize, since Type A matter does not interact with extreme physical heat, but religion says they will run and hide. They will wish mountains could fall on them, which is impractical, because the earth is melting and being technologically reorganized.

The science of the Mount of transfiguration shows that people like Moses and Elijah could be changed from mortal to an enhanced technological state, Type D matter, and live for hundreds of years with a physical body. Peter, James, and John, witnessed that these two individuals had been gone for centuries, and they returned as men, not spirits.[19] This technology is used for everyone at Judgment Day who isn't on the hell list. All these mortals are changed temporarily to Type D matter, so they are not consumed by the physical heat. The process includes being infused with Type A or C light, not photons, and had occurred to Jesus: *the fashion of [Jesus'] countenance was altered, and his raiment was white and glistering."*[20] The process also occurred to Moses when he went up into Mount Sinai.[21]

Hell is a technological confinement of Satan, his followers, and everyone who died on the hell list for a thousand years. In one sense, hell hasn't started yet. It actually starts at Judgment Day. In another sense, hell includes unmitigated remorse for violations of the universal moral code, programmed into us as conscience. Type A matter beings have a perfect memory of their sin, and a knowledge of how it affected their victims, recorded with a precision beyond mortal. Ex-mortal Type A matter beings have the curse of being around unpleasant Type A beings like Satan, plus not being able to move freely about the earth during a thousand years of hell.

After a thousand years, God permanently removes Satan and his Type A matter followers from earth. He resurrects all remaining ex-mortals, i.e. those who were on the hell list at Judgment Day, and places them for fourteen billion years plus, i.e. eternity, on the spare planet earth. They will be Type C matter, meaning they are immortal, eternal, and can eat mortal fish and honeycomb. The spare planet earth has technology designed to last billions of years. It has identical characteristics to our own planet earth in its pristine condition, meaning a physical planet made of atoms, and all species of mortal plants and animals, in their pristine condition. That means for eternity, even those from the hell list have it better than we do as mortals. Exactly as good a planet, with upgraded, eternal technology for their bodies, so they never get sick or die.

Mortals alive at Judgment Day will meet God, but they will be temporarily upgraded to Type D technology so they don't burn in the heat. They will see that God is a scientist who burns wicked people using physical or other type of heat. They will see God is a scientist who has the ability to temporarily upgrade the entire earth to Type D matter. The plants and animals will be upgraded to Type D technology, which means they won't be mortal, they won't eat each other, and the earth will be more beautiful. At Judgment Day, anyone who died in the past six thousand years or more will be resurrected, meaning a permanent upgrade to Type C matter, and they will stay on the upgraded earth with God.

GOD IS NON-DENOMINATIONAL

The information in the Bible is two thousand years old.

We have received tremendous amounts of information about the nature of God in recent centuries through science.

Near-death experiences with the being of light reflect that the being is non-denominational and non-judgmental. He asks what value we found in our mortal lives, experiences, and relationships, not what world religion we joined.

God as a scientist is non-denominational. He is magnanimous. He is un-obtrusive. He has gone to tremendous lengths using science to hide himself from us. Those who want to believe in him or worship him must do so in faith, without physical proof. Atheists are free to disbelieve him. There are other explanations available for how the earth was formed and how the species were diversified, including evolution.

There are eighty billion people who have lived as mortals on planet earth, according to some estimates.

Of these, there was an extremely small percentage who received religion from the God of Israel. From the days of Moses until John the Revelator, God gave religion to his chosen people through prophets. Around 90 A.D., he stopped.

God is an intelligent and rational person.

God has plans to bless the lives of the eighty billion. He has no plans to send eighty million to heaven and 79,920,000,000 to hell.

Since 30 A.D., information about Jesus Christ has become available to much of the world. The Bible is widely available. At the same time, wise persons in cultures across the world have established religions and a belief in God.

We tend to think of God as dependent on religion, as if without religion God would cease to exist. Instead, God's dealings with the Jews, both in Old and New Testament times, were with a very select or chosen group. As to all others, he has allowed them to believe as they chose. This is consistent with God as a scientist, who is not condemning people morally for not believing in him when they have never heard of him.

Science has tried to use data and information to show that everything in the universe can be explained without God. God is hiding from us, and he wouldn't be doing a very good job of it if he didn't give atheists a way out. The brilliance of the massive amounts of scientific information he has provided is precisely that it leaves alternative explanations, rather than proving conclusively that God exists. It is essential to our freedom on this planet that God hides from us. If we had indisputable knowledge of him, we would be influenced by his expectations, and no longer truly free.

God does not force religion on anyone. That isn't what he is known for. When the Israelites entered Canaan, they weren't looking for converts. God wanted them to stay separate from those cultures.

The entire system of this planet is evidence that God allows us to choose what we want without undue influence from him. He does provide moral guidance and accountability, but it can be attributed as much to his nature as a scientist as it is to his religion. You go to hell if you ignore your conscience, which is scientifically programmed into your Type A matter portion of your being. If you become a liar, thief, or murderer, you will go to hell with scientific precision. God's knowledge is perfected, and his science is perfected. Judgment is like a computer program unparalleled and unsurpassed, in fact infallible.

CHAPTER FOUR

WHAT WE DO IN THE AFTERLIFE

To understand what happens when we die, we go a step beyond near-death experiences. We do not go to heaven or hell when we die. We go to the vast world of spirits. The shocking thing for most of us will be that it isn't a mythical heaven or hell, it is a technological condition. There are up to eighty billion ex-mortals there.

Heaven and hell begin at Judgment Day. At Judgment Day, those scientifically on the hell list are burned physically when planet earth melts with heat. The earth becomes a transfigured planet of Type D material. Those not on the hell list are transfigured, temporarily made non-mortal to endure the heat without dying.

What happens to us all if we die before Judgment Day is that we enter the world of spirits. That in itself is perhaps a misnomer, because the world of spirits is on this earth.

TYPE A OR SPIRIT MATTER

The scientific nature of Type A or spirit matter is that it is a prototype of what a physical person, planet, plant or animal will look like. The Type A matter prototype is placed within the physical matter. The Type A is as much an instruction format or program for our physical appearance as DNA itself is. For a person, the soul and the mortal body are one, made of two types of matter. The earth has a Type A matter prototype or spirit, which aided in its creation in the same way DNA aids in the creation of our mortal bodies. Type A matter contains instruction codes. For animals, they contain instincts, migratory flight patterns, and protecting and nurturing instructions for raising progeny.

Everything alive on planet earth, including plants and animals, has a spirit within it. The planet itself has a spirit, or Type A matter prototype, within it.

As a result, when we die, we do not see a different earth. We do not see a spiritual planet. We do not see gardens of beautiful spiritual flowers that are different than the ones mortals see. The earth itself looks the same.

What is different when we die is that we see the spirits of those mortals who have died before us. There are eighty billion of them. In addition, there are forty billion followers of Satan. Satan's followers look the same as those who have died. They are all made of Type A matter.

CLOTHING

You will wear a white robe in the afterlife. You won't be able to take it off. It is programmed as a part of your Type A matter body and cannot be removed. That is why it was a big deal for Adam and Eve to be naked: such a thing was not possible for Type A matter beings. Satan and his followers are scientifically composed of the same matter, and they also wear white robes. They do not get to change clothing or styles.

FOOD

Type A matter beings are not capable of eating food. The spirit plants that they might consider eating are bound within physical plants. Type A beings cannot impair or disturb them. In short, for the thousands of years that some have spent in the spirit world, they have been without food. They have a need for nourishment, but to us it would be an emotional need. They are nourished by spiritual light and by love. Spiritual or Type A matter light is different from ours.

PROPERTY

No one among the Type A matter beings owns any property. They can't build a house, because they can't chop down trees. The spirit of trees is within the living tree, and Type A matter beings cannot do anything to it. There are no watches, because spirits cannot obtain ore from the earth. They don't have some special matter they can access, a spiritual cement, to pour houses out of.

There is no theft among Type A matter beings. God cures thievery ultimately by depriving us of our mortal bodies, which allow us to steal physical objects from each other.

SINS

You cannot commit adultery after you die, because you can't get your robe off.

You can't worship idols, unless you worship the ones mortals have created.

You can't steal.

You can't kill other Type A matter beings.

You cannot assault other people. You can't touch mortals, and when you touch other Type A matter beings they will feel no pain or impact from your violence. If you punch someone in the nose, your fist will stop where his or her nose begins. There is no physical impact.

You can commit adultery in your thoughts, but you will find it profoundly unsatisfactory.

The one thing you can still do is lie, especially to yourself. It does get harder though. There is an indication that others perceive your thoughts, because you are no longer encased in a mortal body.

HOUSES AND LIBRARIES

The only people who have houses and libraries on this planet right now are mortals. Some have seen visions of houses and libraries in heaven; well and good. They are seeing visions of the future. Heaven includes mansions, houses, libraries, and beautiful buildings. These things are for mortals or resurrected, Type C matter beings, who are physical beings. Many things can be made of mortal and resurrected matter, due their scientifically physical nature. These things will be a part of heaven after Judgment Day, not before. If you want to read a book in the afterlife, you will need to find a mortal who is reading, and read over his or her shoulder.

MORTALS HAVE A GREAT ADVANTAGE

Most people who have a near-death experience think that the afterlife is preferable to their mortal life, and they aren't anxious to go back. They describe the unpleasant, heavy nature of a mortal body, particularly when it is deathly ill or dying from injury.

Feeding the hungry and clothing the naked are for mortals. Playing football and baseball are for mortals. Living life to its physical fullest is for mortals. In the afterlife, you can watch mortals do anything and everything.

Spirit prison refers to the nature of being a Type A matter being: you are unable to engage in physical activities. After a lifetime of being able to do physical things, it can feel like prison.

The afterlife is all see and no do. This means regarding physical activities. There are certainly things to do in the afterlife, in spite of the limitations.

SOCIALITY AND INTELLECTUAL PURSUIT

The eighty billion people who have died on this planet

live in the groups of people they want to be around. They live with neighbors, fellow church members, and tribe members. They can associate with anyone they want, but they tend to associate with their own culture and people.

If there are 10,000 mortal Christian churches, there can be that many or more in the afterlife. They can meet once a week or daily. Elks and Rotarians and bridge clubs have as much opportunity to gather as ever they did, although the bridge club will be without cards, and the Elks will have no alcohol.

RELIGION AND SCIENCE IN THE AFTERLIFE

Religion and science unite to a degree in the afterlife. Many will have met the being of light, who is non-denominational and nonjudgmental. The atheistic scientist who sees the being of light will know that there is such a being, a non-malevolent, loving being inquiring about how much value that person found in life. Everyone who dies experiences the reality of Type A matter and all its scientific characteristics.

TYPE A MATTER CHARACTERISTICS

The information and data show that we are able to move as Type A matter beings without walking or running. We are able to move like thought. There remains a speed limit on Type A matter, to prevent persons from this planet from venturing into space. Stars are placed light years away; the nearest would require four years of travel.

We have GPS. Our GPS is a poor mimicry of navigating and locating systems available to us as spirits. We will have access to the globe. We can observe mortal countries, cities, and cultures to the utmost. You can spend a year as a tourist, or a hundred years.

THE GREAT DEBRIEFING IN THE SKY

We assume that when we die, we will have a near-death experience, without the near part. More importantly, we assume someone, divinity or our favorite family member, is going to tell us what in heaven's name, or perhaps what the blazes, is going on.

That doesn't happen the moment we die. The great debriefing happens at Judgment Day. That is when we all meet our maker. If we believe in God, that is the time when we will find out who he is. If we believe in Jesus Christ, that is the time to witness firsthand that he is God.

God has created us as mortals on planet earth as the first step in an eternity of being. We spend, on average, eighty years as mortals, and we'll spend fourteen billion plus years after that. Part 1 of this process is a mortal planet. The rest of the parts happen after we die. They include a place where people gather after they die. We call it the afterlife. It is a vast world of spirits, and it is located on this earth. There will come a Judgment Day, and that is the day for the great debriefing. It is the day when everyone will know who God is.

We know all about the mortal part. We want to know about the part that happens after we die, but before we are resurrected. This is the spirit world, keeping in mind that spirits are people, in the sense that we still have our identity after we die. We still have our friends and family. We have the same social networks we've always had, with a different type of technology for daily life. There will come a day of resurrection. After the spirit world comes the Judgment Day and the resurrection. We will all be resurrected, and we will spend the next fourteen billion years plus in the place we choose to be. We choose where to be by our actions in mortal life and in the spirit world. Most of our actions have to be done in the mortal world, because we are limited in the amount of good we can do in the spirit world.

THE PATTERN FOR MORTALS

The nature of God's dealings with mortals in Judeo-Christian religion is that of limited information. The stark example is that neither Jews nor Christians have received new scripture in the past nearly two millennia. Christians use most of the books taken from the Jewish Torah, without the oral traditions, as their Old Testament. Christians picked which books to include. In the King James version, the title page of the New Testament is clearly written as: "The New Testament of Our Lord and Savior Jesus Christ." The word Testament means witness. The Old Testament doesn't have a similar title page referring to Jesus Christ. Presumably, it is the "Old Testament of God".

In general, Christianity operates under the presumption that we as mortals did not exist prior to being either conceived or quickened in the womb. Is this also true of Jesus Christ? Under a nearly unanimous Christian theological viewpoint, Jesus Christ also began to exist at the time of either his conception or quickening.

The lack of information, in spite of the length of the Bible, leaves us with confusion and contention among Christian religions over things as fundamental as the nature of the Godhead.

There are a variety of world religions, and each serves God's purpose for that people.

THE PATTERN CONTINUES FOR HEAVEN AND HELL

God's pattern of giving us limited information is especially evident for heaven and hell.

In the Jewish religion, there is a vacuum of information about the afterlife, so much so that it is debatable whether Jews believe in an afterlife, although Saul sought a witch to bring back Samuel from beyond the grave.[22] There is some acceptance that there is a resurrection.[23] The fact of a resurrection implies an afterlife, but with no detail.

The Christian notions of heaven and hell come primarily from the New Testament, with the Book of Revelation giving the best description.

It is fair to say that the notions of heaven and hell for Christians are filled with controversy.

I AM NOT A CHURCH

I delight in the fact that I am free to ignore all churches in writing this book. Churches are bound by centuries of ideas and culture.

In my view, we have a great deal more information about heaven and hell than anyone did 1,500 years ago, or even 100 years ago. Ignoring churches, temporarily of course, allows us to consider all available sources of knowledge, and to combine all the data into a reasoned view.

The information we have includes the Bible, scientific evidence, and near-death experiences.

Every effort will be made to present the information in a non-denominational way.

GOD'S IDEA

God is the type of person who created mortals so that we could be eternally happy. He wants as many of us to be happy as possible. Part of this book talks about why having evil in the mortal world results in greater happiness.

God wants to send the greatest number of people he can to heaven, and the least number to hell. To say that God intends to send everyone who does not belong to a particular church to hell does not given God any credit for having love and common sense. The vast majority of people who have lived on this earth never heard of Jesus Christ during their lives. It is impossible for me to believe that God intended them all to have, on average, eighty years in mortality followed by an eternity in hell.

GOD'S HELL LIST

The hell list is found in the Bible. The hell list is for people who are liars, thieves, sorcerers, whoremongers, and murderers when they die.[24] It isn't for all of us who will still be sinning when we die, because we all will be. We might even tell a lie right before we die, but death-bed lying doesn't count. It doesn't get us sent to hell. This list isn't for people who committed adultery once and then repented. By repent, let's say they never do it again; everyone knows about it; and they have remorse.

The litmus test for the hell list is that a person, at the time of his or her death, has become the type of person who lies, cheats, steals, or harms others without remorse. They reach a point where they generally don't feel guilty anymore. Their conscience is numb. This is a bad sign for anyone who doesn't want to be on the hell list.

"BELIEVER"

Part of the problem mere mortals have is that the "unbelieving" are included on the hell list.[25] We've spent much time and energy in contention over people who don't believe in the same church as us.

Catholics, Protestants, Evangelical Christians, and Jews will all go to heaven, as long as they aren't the hell list. So will everyone else who isn't in one of those churches, as long as they aren't on the hell list.

The reason churches have a tendency to think members of other churches are going to hell is in Matthew 7:21-23:

Lord, Lord, have we not prophesied in thy name? and in thy name have cast out devils? And in thy name done many wonderful works? And then will I profess unto them, I never knew you: depart from me, ye that work iniquity.

I draw a distinction between the people I've know from all these churches who are sincere believers, and those who are working iniquity. I have known many Catholics and Evangelical Christians who obviously have a deep love for Christ, and I can't for a moment think that one group or the other is going to hell for believing in the wrong church. To me, that verse can be fairly explained by saying that if you become a liar, thief, or murderer being in a church isn't going to save you.

The pattern God has followed for mortals confirms the notion. Common sense tells us that the reason that God didn't give us more scriptures in the past two thousand years is that he intended to save as many of us as possible, and that he provided a way to do that in spite of the large number of different churches.

From the Christian standpoint, it was enough that in 33 A.D. a very small percentage of the world's population knew who Jesus Christ was, and today billions know. A gracious view of Jesus Christ is that he atoned for every mortal who ever lived, not just those who were taught during mortality to believe in him. The only way they go to hell is if they don't repent. This implies that they have to be taught how to repent, they choose not to, and they become the kind of people on the hell list.

CONSCIENCE, THE UNIVERSAL TEACHER

Imagine a tribal chief from 1,300 A.D. in America who leads his people as best he can. He never hears of Jesus Christ during his lifetime. He is the type of person who feels guilty when he steals, and it changes his behavior. He teaches his people not to lie or steal. They are good people, because they listen to the universal teacher, conscience. Members of a neighboring tribe have a different kind of chief, and they become thieves and murderers. They ignore their consciences for so long they start teaching their children that murder is a sign of strength and bravery. They live in warfare.

If you believe Christ atoned for sin, you understand repentance.

If you believe in feeling remorse and choosing to live as good people, you are living by the same standard.

Describing heaven and hell requires us to consider what happens to these two chiefs as mortals, and then in the afterlife. It also requires an understanding of Judgment Day, or the end of the world. Much of this has to do with timing, as we will see in later chapters.

NEAR-DEATH EXPERIENCES

Near-death experiences are the best source we have outside the Bible for the fact that the afterlife is reality, not a dream-like world. Those who have had near-death experiences feel their existence as spirits is perfectly natural, a transition from one state to another. They are greeted by loved ones. At the point of death, or separation of their spirits from their bodies, they often see their mortal bodies and the doctors and nurses who are trying to revive them. Some of them see a being of light they identify as divine.

What doesn't happen is that someone explains to them the doctrine of the trinity, baptism, being born again, or any other of the many doctrines Christians contend over. They may ask "What's going on?" but they don't get an answer. We aren't going to get that answer when we die either; not until the Second Coming.

THE LIFE'S REVIEW

We think of the life's review as described in near-death

experiences as a spiritual experience. It is spiritual, in the sense that it happens to spirits, i.e., to those who have had their spirits separated from their mortal bodies, albeit temporarily. We might better describe it as real and technological. One of the characteristics of spirits is that they are not impeded by a mortal body.

A mortal body in some ways has a tremendous negative impact on our normal characteristics as spirits. We get tired and our minds get fuzzy. A drug as simple as alcohol can cause our minds to malfunction.

Our mortal bodies are well worth the sacrifice of spiritual mental acuity. Eighty years of experience with good and evil is so valuable that God uses it to start off our eternity.

The memory capacity of a spirit is unimpeded. Put another way, our spirit is a recording device. Our spirits record everything about our lives, in infinite detail. Our brains record relatively little.

Those who have near-death experiences realize the life's review seems natural, vivid and real. It is not like a religious tutorial on life designed by God. It is natural because our spiritual memories return to us once we are free from our bodies.

In other words, a life's review doesn't prove God exists.

Seeing relatives doesn't prove God exists.

Some have described seeing a being of light. He doesn't tell them to make sure they join a particular church as soon as they get back to their mortal life.

We still have to live on faith after we die, up until Christ's Second Coming. No one gives us definitive religious answers the moment we die.

This shows us the pattern by which God deals with his children. He sends prophets and reveals scriptures to those mortals who seek religion. His interaction after we die is to send us a being of light whose purpose is to cause us to reflect on our mortal lives.

SPIRIT MATTER

Science is for mortals. Our interaction with physical matter allows us to move matter. We can build machines and run experiments.

The nature of spirit matter, as implied by religion, is that it does not allow us to move or build things.

LIFE AFTER DEATH

God designed mortal life to be full of valuable experiences for us all. Both good and bad experiences help us to grow, become more resilient, and to learn. We become better, stronger, kinder, and wiser.

God designed life after death in a very similar way.

As mortals, we live by faith. Nobody on earth can tell us which of thousands of Christian churches to join. We try to find one we feel comfortable with; we may feel it doesn't matter which of the many Christian churches we attend. The reality is that persons in all of the Christian churches are judged by the same standard: If they aren't on the hell list, they aren't going to hell.

God uses the technology of the second type of matter, spirit matter, to control our circumstances after we die. Life in the spirit world is different from mortality in a crucial way: There aren't very many sins we can commit after we die.

THE VAST WORLD OF SPIRITS

The key to understanding what eighty billion people have been doing since they died is to realize that they do not leave earth. They remain among mortals, unseen. They can see and observe everything mortals are doing. They see the same earth and mortal buildings that mortals see. They can travel anywhere they want on earth, gravity notwithstanding. Spirit matter is not made of photons, but it has the same speed limit. Spirits are confined to the vicinity of this earth and solar system because the nearest star is 4 light years away. God has not placed a barrier anywhere preventing spirits from leaving earth, he just made the distance a deterrent.

All living physical things have spirit or life force in them, which keeps them alive and growing. The spirits of humans are confined within the mortal body. The spirits of animals are confined within the animals' bodies. The spirit of a tree is confined within the tree. The result is that spirits see mortals who have died, and aside from that they see the mortals' physical world and buildings.

A PLACE OF WHITE LIGHT

There is a possibility that those who have died are able to see a place of white light, and to focus on that place and avoid seeing mortals and the mortal earth, or to see both.

WHAT COMES NEXT

We want to know what comes next. We want to know what happens when we die.

We expect that we will go through the dark tunnel, towards the light. We see the being of light, and experience our life review.

At this point near-death experiences don't tell us anything more about what is going to happen, because this is the point where the person gets sent back to his or her body.

This time we aren't getting sent back.

WHAT WE DO AFTER WE DIE

Once you are done seeing the being of light, you will be on your own as a spirit on a mortal planet. I don't mean completely on your own. You will have lots of friends and family. If any of them are acting as your guide, they can explain the way it works to you to be a spirit on a mortal earth.

You'll be a spirit being, among many spirits, remaining unseen by mortals, but seeing everything that mortals can see.

In near-death experiences, people sometimes describe seeing family members who are spirits as being translucent. When Jesus appeared to his apostles, they thought he was a spirit even though he was a resurrected being.[26] This suggests that if a spirit appears to a mortal, the spirit will look like a solid person, surrounded by or emitting light, perhaps, but not translucent. I've always assumed that spirits look that way to each other as well: they look like solid people, like normal people only without imperfections, and with skin and hair of normal color, although infused with light. To simplify, I think that they look to each other like normal people in white robes look to us as mortals, except that spirits naturally radiate light. It is not photons; we understand that type of light. It is a technology of light associated with spirit matter, and other spirits see it, even though mortals cannot.

There are other possibilities, one of which is that the many accounts from near-death experiences are accurate in describing that to other spirits, spirits look translucent, or like a ball of light, or a mist, and not like a mortal person. The type of daily experience they have wouldn't change from what I describe, except that the visual aspect of seeing many other spirits around would depend on their appearance. That is only partially impossible for us to comprehend or imagine, which means it is partially possible.

If what you want when you die is to rest, feel free. The planet is yours. Go see the Grand Canyon, or any other beautiful place, for as long as you like. You won't be sitting on a cloud, or in a vast white area filled with light, with nothing to do. You will be on planet earth, and you can see the earth just as you always have.

When I say go see the Grand Canyon, I don't think you

will have to walk there. The descriptions we have of the technological properties of spirits is that they are able to travel like light or thought. Gabriel traveled from the presence of God.[27]

It is not possible for spirits to build using spiritual matter. The technology of spirit matter is that it is confined to living forms—plants, animals, humans, or the earth itself.

To put it in practical terms, you can see the world, as a tourist, after you die.

For spirits, there is no eating.[28] There is no getting fat. There is no exercise, and no physical exertion. These things are characteristics of a mortal body.

Our time as spirits will be spent in social gatherings or interactions with friends or loved ones. We can attend with the same people we did before; we can attend the same social clubs, or meet with friends. We can attend the same church congregation we always have, unseen, or we can gather with other members of the congregation who have died and have our own meetings.

We can also watch the mortals living. We can watch them play football games, anytime, anywhere, as many as we want. We can watch them race cars. We can attend their school and university classes, or we can hold our own among the spirits we know.

There are ample opportunities to learn and to associate. There are no opportunities to feed the hungry or to visit the sick. If we missed those opportunities in mortality, it will be too late.

The good news is that after we die, we will no longer be driven by hunger or other need. We will have no motive to steal. We will have no physical drives.

The bad news is that we become relegated to watching and observing mortals, or ignoring them. We will still be as emotionally needy as spirits as we are as mortals. We will still crave attention. We will still feel guilt for things we did wrong, if we did not change before we died.

The parameter is that God does not unnaturally interfere with spirits after death any more than he does with mortals. He doesn't provide computer access to vast databases of the cosmos; he doesn't build large libraries out of spirit matter for spirits to read.

Peter is not going to be there for you to ask your definitive religious questions. Instead, he'll be one of your judges at the Second Coming.

Those inclined to share religion may do so: *For for this cause was the gospel preached also to them that are dead, that they might be judged according to men in the flesh, but live according to God in the spirit.*[29] Preaching wouldn't be necessary if we were all in the presence of Christ, learning from him directly, after we die.

Some have seen buildings, not constructed by mortals, in near-death experiences; in my view, invariably they were viewing future buildings.

The nature of a spirit is that he or she cannot move physical matter, nor can the spirit modify or move other spirits or living things. Spirits can pass through human walls or buildings, but not through a living tree, plant, or animal. Reported cases of possession of a mortal body by an evil spirit are abnormal systemic malfunctions, technologically possible but a variation from normal rules.

It is not possible for spirits to change clothes. The white robes are more than mandatory, they are natural. If a spirit looks to other spirits like a translucent ball or mist of light instead of a person, that would explain why they don't change clothes. There are instances in near-death experiences where the experiencer sees the spirit of a family member wearing the mortal clothes he or she wore before they died; to me, this is a function of God controlling the information he provides in the experience. The experience is still real and technological.

THE PURPOSE OF LIFE

One of the reasons we experience mortality, with both good and evil, is because it allows us to become. We become wiser, stronger, kinder, and more knowledgeable, all through our experiences.

Another purpose of life is to find out what we really want. We have eighty years of mortality to find out for ourselves how we want to spend eternity; this is followed for the vast majority by a much longer period in the spirit world. For example, the average person prior to Christ's birth spent 50 years in mortality, and has spent more than two thousand years in the spirit world.

Our time in the spirit world will allow us to observe mortal actions, and the consequences. It will be obvious how much mortals can accomplish by serving and helping each other, whereas for spirits after death, there is much less that can be done to help others.

BEGINNING ETERNITY

We spend about eighty years as mortals, and perhaps thousands of years after that as spirits, depending on whether we are mortals who die in 3,500 B.C. or 2050 A.D.

Then we live on for eternity.

God chose mortality and the spirit world as a means for us to begin eternity by becoming the type of beings we want to be.

There are at least two types of resurrected beings we are aware of: Those who want to be with Christ eternally, and those who don't.

Those who want to be with Christ will be on earth with

him: *and they lived and reigned with Christ a thousand years.*[30] This is followed by a final battle with Satan, and then they will spend an eternity with Christ. They will be people in white robes. They will be able to eat food. They will be resurrected beings, meaning they will never get sick or suffer. There will be no enmity among animals. The vast majority of people who have lived are not liars and adulterers at the time they die. The vast majority of people who have lived on earth have never had the opportunity to believe in Christ. They aren't going to get sent to hell for not having heard of him during their mortal lifetimes.

The group who are on the hell list are going to be cast into hell with Satan and his followers for a thousand years. After that, at the final judgment, they will be resurrected.

The promise is that they will be placed on an earth like the one we have now, characterized by the physical beauty of a world made of atoms and molecules, with enmity among animals.

Imagine being a resurrected being, with no sickness or death, and no human pollution, living eternally on a pristine earth duplicating this one. There will be lions and tigers and bears, and these carnivores will be eating little lambs and small zebras. They won't be eating the resurrected people, because resurrected beings are immortal and indestructible. There will be physical animals and physical trees, made of atoms.

We pray for peace on earth, with food for all. Resurrected beings from the hell list, after being delivered from hell and saved by Christ, will spend eternity with peace on earth, with no hunger or suffering. When I say no hunger, I'm not saying they won't enjoy eating. They will. Being saved means being resurrected; it also means having Christ relieve you from feeling any further guilt for the rest of eternity.

Resurrected beings can eat fish and honeycomb.[31] I hope

they can also eat roast pork cooked over mesquite coals, with a big turkey dinner every Thanksgiving. That would be on a planet like the earth is now, i.e., made from atoms and molecules and physical technology. It won't work on the earth where Christ will dwell, because it is a higher level of resurrection technology for all living things. I'd like to believe those with Christ will have better food, and not just manna. Planet earth ultimately becomes like unto a sea of glass and fire, a resurrected earth, not corruptible, and indestructible.[32]

That doesn't mean those who were on the hell list will have the opportunity to keep sinning for eternity; they will be on a planet like this one, where they eat, drink, laugh, and sing. They won't steal, because everyone will have white robes and plenty of food. They won't murder. They won't commit adultery, because they won't be able to take off their robes. More than that, accepting a resurrected body is a form of contract that they will be incapable of sin thereafter, eternally. They won't be able to drive cars, because manufacturing cars and other machines impairs the earth, and cars pollute it. They won't need transportation, because resurrected beings have increased capacity for travel.

ETERNITY

Christ already made plans for all humans. He already decided to save them all. He died for all their sins. He will resurrect them all.[33] The only ones he won't forgive, at the final judgment, are the sons of perdition, like Judas Iscariot. Everyone else will know who Christ is, and accept his atonement and forgiveness; every knee shall bow.[34] The remarkable thing about the sons of perdition is that they will know who he is, and still refuse his forgiveness.

The benefit for Christians is the blessing of knowing Christ and receiving forgiveness in mortality, and the blessing of having all the knowledge and guidance of the Bible.

Everyone who isn't on the hell list and wants to spend an eternity with Christ has the opportunity to do so after they are resurrected.

This is why our contentions over doctrines among the Christian churches must be so offensive to Christ, the one who died to bring all of us back to him.

CHAPTER FIVE

ETERNITY TRANSCENDS RELIGION

When I was twelve, I began looking for religious answers. I went to church regularly, but that wasn't enough. It wasn't that I didn't believe what my church taught, it was that I wanted more. I was not willing to take my church's word for it. I wanted a direct source. I also visited other churches with my friends of other faiths.

I read the Bible cover to cover in high school, and was fascinated by it, not just in a religious way, but in a bigger way. It wasn't just about religion, it was about life. There was the contradictory information that God loves all his children, yet some were chosen. There were regular plagues and famines, even in Israel, among the chosen ones. The Israelites were conquered and placed in bondage multiple times. There was information about how the ark of the covenant showed a vision of God to the High Priest or prophet.[35] There was a description of what the high priest or prophet saw in the vision. Ezekiel described it: Circles of enveloping fire, comprising a blazing gate.[36] Beyond the gate was a city, with a throne upon which God sat, and the mercy seat on the ark of the covenant is fashioned after the throne; the holy of holies was fashioned after the pattern of what the prophets saw when they looked into heaven.[37] It is difficult to tell whether the prophet is describing past or future; if the city is New Jerusalem, it is future.[38]

I was specifically looking for an answer from God instead of from a church.

It took me five years to get an answer. Better stated, I received lots of answers; it took me five years to understand how the process works. I was very persistent. At times, I felt anger and resentment toward God because I believed he was making it difficult on purpose. Many people stop trying.

CONTROLLED KNOWLEDGE

How does God teach us? Why does he hide himself from us? How can he create a planet for us, without us knowing whether he did it in millions of years or in seven days, or without us being able to prove he even did it? If we all knew we wouldn't be arguing about it.

God controls the knowledge he gives us. The world has the Bible. It is two thousand years old. God forces us to act on faith by limiting the information he gives us. The Bible is full of valuable information, and we are free to believe it. We are also free to disbelieve it.

We don't remember anything before our birth. That might be because we didn't exist before birth. It might be because God controls knowledge by giving us amnesia.

I have been angry at God sometimes for not giving me important information. I was angry at him because it took me five years to get an answer from him. I believed he was doing it on purpose, and I felt manipulated.

God is not doing anything wrong, however. Morality does not require God to give us all the information he has at once, or ever. He owns the information.

I'm glad he is sharing it.

WHAT WE KNOW

There are two types of knowledge: physical or mortal knowledge, and spiritual knowledge.

MORTAL KNOWLEDGE

We know we are on a planet. We know where we came from physically: we were born as babies. We know we are mortal. We know we die. We know there are thousands of churches and several major world religions.

We know more about ourselves and the world than we did two thousand years ago.

Science tells us about an astounding array of physical matter, in astounding depth. We know about the cosmos, galaxies, solar systems, and planets. We know about continents, countries, and cities. We know about human bodies, organs, and cells. We know about molecules, elements, and atoms, all the way down to quarks.

Science is knowledge of the characteristics of physical matter.

SPIRITUAL KNOWLEDGE

First, we have spiritual knowledge based in personal experience. We know we love each other. We know sometimes we hate each other. This is accepted by knowledge as science. We consider it part of physical reality. It isn't.

Emotions of love, hate, and every other type are spiritual, not physical.

A physical body dies. There is no love in a dead body.

Still, we can't prove that our emotions are spiritual, not physical. God remains hidden if we don't believe emotions are spiritual. Atheism remains possible.

Second, we have spiritual knowledge based in religion, and by that I mean the Bible, not churches.

Religion describes that we are spirit children of God. It teaches that our spirits dwell in our mortal bodies, and that the spirit separates from the body at death.[39] It describes angels, such as Gabriel, who appeared to Mary, the mother of

Jesus.[40] The Bible is two thousand years old so that we can choose to disbelieve it.

Third, we have spiritual knowledge by the spirit of God. Do religious believers know God lives? Some do. Others just believe.

When I was twelve, I looked for an answer from God. It took five years to get the answer I wanted. Here is what it feels like: Go to a memorial service for a soldier who has died for his or her country, and watch the love his or her spouse and small children have. You will weep with them, and share their grief. Grief is a spiritual experience. Grief is the spirit of God, which causes you to weep as though your heart is broken, as though you will suffocate. If you have a broken heart, and reach out to God, you feel the power of his spirit.

Instead of at a memorial service, maybe you will be telling God how sorry you are for something you did. Maybe you will be telling him how much you love your children, and that you don't want them to be hurt by your financial problems, or by something else that's happened in your life. If you feel his power in response, you will weep with a mix of sadness and joy. That is how people find out God lives. That is how they know prayer works. That is how they know God loves them.

People from every religion and no religion can feel it. Some feel it and do not accept it as a manifestation of God. Others feel it, and they know God lives. It is designed to be spiritual knowledge, not physical knowledge.

UNDERSTANDING SPIRITUAL THINGS

So many of our great books, plays, and movies move us to tears using spirituality. The heights and depths of the emotions and experiences of the characters mirror events in our own lives, and tug at our heart-strings. At times, they wrench them.

We have vast knowledge about our physical world, where

we'll be for eighty years as mortals. We know very little about heaven and hell, or the billions of years ahead of us.

Spiritual knowledge is important to us. It is precious. We search for it. I've spent my life searching for it.

We believe more than we know.

We believe our spirit is within our mortal body, and it will leave our mortal body when we die.

LIFE IS THE ANIMATOR OF PHYSCIAL MATTER

The mind is worth more than the body.

We know what a mind is. It is us. It is our identity. We feel it physically somewhere in our head, and it is connected to our physical eyesight, as we think and feel emotion daily.

Our minds are life.

Without our minds, our bodies are mere shells. All of the technology and parts are still there, but they are not animated.

Most of what we are is in our minds, not in our bodies. Our emotions and feelings are more important to us than physical sensations. The love we have for our spouse means more than a hug, although the hug is wonderful.

We understand how a TV is a mere panel of plastic without being connected to electricity and a broadcast source. Our body is the same way.

Life in mortal bodies is similar to electricity in mechanical objects. You know if your battery is dead, because your car won't start. You have three thousand pounds of physical matter at your disposal, powered by the strength of 200 horses, if only your battery weren't dead.

Your body is far superior to any machine we have ever created, and you would be welcome to use it, if only your mind was still connected to your body, if only you weren't dead.

We know what a mind is. It thinks and feels emotion.

We know what a body is. It is a technological marvel

surpassing any machine, unless the mind is missing, in which case we want to put it somewhere we won't observe its decay.

THE SPIRITUAL CONNECTION TO OUR BODIES

The secret of life is how to connect life force to physical matter.

The first part is to put life in the cells. It works for plants and animals, as well as humans. Both a plant and a person have life diffused throughout the cells. A plant doesn't have an obvious overall controlling mind as its identity, like an animal or human has. The instructions which make plants grow are partly in the DNA in each cell of the plant, but cells are alive on their own. They die if they are disconnected from the plant. The DNA is still there, just as it is in a clipped strand of hair, but the life quickly leaves.

The second part is to connect a mind, which is the life force, to a physical body.

If you want to move your finger, you just think it, and your finger moves. You have mind control over the 170 pounds of physical matter in your body. Your mind, the force of life, is interconnected to your brain in such a way that your mind thinks a thought and your brain fires an electric pulse through your neurons to your finger.

Breaking your finger causes physical pain.

Being depressed is painful, but it isn't physical, it is spiritual.

FOUR TYPES OF PEOPLE

Religions tells us four types of beings exist: mortals, spirits, transfigured beings, and resurrected beings.

Why didn't God just create us as resurrected beings from the start? In fairytales, we talk about happily ever after. God had that option. If all we really want out of existence is to have eternal bliss in the presence of Jesus Christ, what is the point of being victims of divorce, cancer, or murder?

GOD'S TECHNOLOGY

The four types of people represent four types or paradigms of technology. When I say technology, I am not talking about machines, but bioengineering, or living technologies. Our machines are crude efforts to mimic God's biotechnologies. God designed a pump called the human heart. We try to imitate the design of the heart with mechanical pumps. The human eye is superior to any camera we've ever built; it is a better technology than a machine, and it is alive. It is programmed to be self-repairing, like everything in our body. Cut your finger, wait two weeks, and it is healed.

GOD IS THE GREAT SCIENTIST

The purpose of mortal science is to help us understand how God does things. He has left a record in the earth of most of the things he's done; the levels and strata in the earth record the data. Microwave data record the formation of the universe because God wanted to leave us a record of what he had done.

The world can function without scientific knowledge. Two thousand years ago people lived an agrarian life. They built houses and cities and they raised families. They had wagons with wheels, a type of technology. They didn't know that the movement of tectonic plates in the earth's crust causes earthquakes. They didn't know about atoms and molecules.

God gave us science when he felt we needed it. He used inspiration to provide scientific knowledge at the right times in history. He didn't provide nuclear weapons in two thousand B.C. for a reason: on average, it has taken 126.2 years for humans to blow themselves up once they obtain nukes. That's assuming there are other planets he can use for comparison.

We won't assume that. Instead, let's just say it was prudent of God to provide us the inspiration to create nuclear weapons towards the end of our mortal world's life cycle.

God's purpose for science is to show us how he does things. Science shows God's handiwork. We build fine harps and violins. God builds planets.

FOUR TYPES OF MATTER

We've discussed four types of people: mortals, spirits, transfigured beings, and resurrected beings.

The four types of beings represent four types of matter: physical or atomic matter, spirit matter, and resurrected being matter.

We know a great deal about physical matter, and a few things about spirit and resurrected matter.

Eighty percent of the universe is made of dark matter. I don't know whether dark matter includes spirit matter and resurrected matter. I do know that religion has taught for two thousand years or more that there are other types of beings than mortals, and science has finally shown that we have no scientific understanding of eighty percent of the matter in the universe.

THE PLAN

God uses eighty years of mortality to start off our eternal existence, and he let Adam and Eve near a tree which allowed evil in the world.

His plan was to combine use of the vast technologies of mortal bodies and a physical planet with the tremendous emotional and imaginative capabilities of the mind.

THE ASTOUNDING VIRTUOSITY OF THE MIND AND PHYSICAL BODY COMBINED

Your mind makes a supercomputer look like a sissy.

I understand a supercomputer can beat a master—at chess. Let's see the computer operate and function a human body, the way a mind does. Let's see a supercomputer heal a paper cut.

Our minds keep our balance every time we walk or run, integrating enormous amounts of data instantaneously. It takes about a year for a baby to learn to walk. Not only are the child's muscles growing, but the brain is programming itself through trial and error to achieve the highly complex task of walking.

Our minds process staggering amounts of data from visual input. Drive a mile, looking at the trees and houses around you. If you have an idea of the number of pixels it would take to get high definition video of what you see as you drive, consider the amount of data your mind processes from visual input alone, all day long every day.

Your mind processes the data to determine what is most important. Threats, like another car crossing the median line, are processed with priority.

In addition to the visual input, our minds are processing touch, smell, sound, and taste.

On the issue of healing a papercut, how does your mind do that, when you aren't smart enough to do it yourself? Your body heals your papercut. The cells act the way they are supposed to, duplicating and healing. God designed the human body as a self-repairing biotechnology.

The human body is extraordinarily high tech, operating continuously in ways we don't understand, like healing itself. Our greatest ignorance is about the key of life itself. God knows how to make a person alive. The rest of us are still trying to figure it out.

The mind also stores knowledge. It does so selectively. It processes that which is most vital to remember, and forgets the rest of the input and data. You have a lifetime of knowledge about language, diseases, friends and family, past experiences, places, and skills. Your mind stores all of it, and when you die, you will be able to access all the data. Right now, your mortal body blocks some of the data.

OUR SPIRITUAL CAPACITY FAR EXCEEDS OUR CAPACITY AS MORTALS

On your best day as a mortal, you will use ten or twenty percent of your mortal brain.

If you get drunk, you'll use a lot less than that.

When you die, you will understand the full capacity of the human mind. Your life will flash before your eyes, because everything your mind has recorded will come back, even though your mortal body couldn't keep up with the information.

INTEGRATION OF MIND AND BODY

The mind and body are integrated almost seamlessly. They say you shiver when someone steps on your grave. That is a spiritual shiver. When you are severely injured, you go into shock, a medical condition. Shock is also a spiritual condition; it is your mind registering the fact that your mind and body can and very well may separate from each other.

Our bodies are full of chemicals that integrate our spiritual needs into our physical ones. We have a spiritual need for acceptance and approval. Little children are obvious in their desperation for attention. We all have it.

Depression is physically manifested by chemical imbalances in the brain. God designed the physical effects of those imbalances to mimic the sadness of our mind. Our minds are extremely sensitive instruments recording vast data, and part of the data they record is how imperfect mortal life is. There is a schism between what our minds expect and what physical reality is. Our spirits don't expect our bodies to be overweight, yet sometimes they are. Our spirits expect order and perfection, and we see chaos in our lives.

Chemicals cause our bodies to mimic all kinds of the virtuosity of emotions our minds are capable of. Alcohol is a chemical which magnifies what we are feeling. Getting high on heroin or cocaine is a physical manifestation interacting with our spirit, producing pleasure in body and spirit which is addictive. Drugs are able to bridge the gap between body and spirit, as well as to mask physical pain.

We sometimes feel spiritual maladjustment in our physical bodies. We sense that we have tremendous capacity as immortal spirits, yet we are stuck in a finite body, constantly making mistakes. We feel the passage of time like a thorn under a fingernail, because we are eternal beings in a body designed to self-destruct in eighty years.

MORTAL TECHNOLOGIES MIMIC GOD'S TECHNOLOGIES

The migration of birds and other animals occurs because of better-than-computer precision of programming of instructions and information into animals as a biotechnology. Life force is used to program information into plants, animals, and humans. We might call it animal instinct. In the same way God showed us that he uses DNA in every cell to program the growth of plants and animals, he also programs animals with information needed for the complexity of life. There is a spirit in an animal, just as there is a spirit in a human, and that spirit can be programmed with information necessary for survival.

Animals come spiritually programmed with directions on how to migrate. We understand how the magnetic field of the earth and the configuration of stars can guide us on sea and land. We use compasses. Compasses are a poor mimicry of spiritual technology, which needs no compass. We understand ants used pheromones to communicate in amazing ways with microscopic amounts of chemical signals.[41] We use cell phones to talk around the world. God uses prayer. We use the internet. God uses a spiritual connectivity between minds. We use GPS and tracking devices. God has something better. I don't understand it; I just know he has had billions of years to perfect the technology we've been working on for a few hundred years. He inspires us with portions of his technology. He tells us the vision of God opens above the ark of the covenant, so a prophet or high priest on earth sees the throne of God.[42] We think God doesn't understand technology. We believe, incorrectly, he doesn't have communications technology far more advanced than cell phones and computers.

GOD-LIKE POWER

God can move matter without touching it, which is a miracle. He seemingly doesn't touch it. The ability is similar to the ability to have a plant function, with all the cells coordinating in a living plant. He organizes a plant using life force to organize the plant cells. He uses life force to move matter.

We should understand how miracles work, such as the bread and fishes, or turning water into wine.[43] It appears to me that God has all the information about a loaf of bread or a fish digitized, and he can reproduce is remotely. He can move a mountain because he has digitization technology which can read the data and reproduce it. We have 3-D printers; he has remote assembly. We can't so much as reproduce a grain of wheat, not even on a 3-D printer; our problem is the complexity of a cell—we aren't quite up to speed on how to get all the DNA and other information in the right places.

Mortality with physical bodies is God's means of sharing his God-like power, moving matter without touching it, with us. By integrating our minds inside physical bodies, God gave us direct mind control over our body weight in physical matter. The integration and use of brain signals through firing neurons masks the fact that our minds control physical matter. Our minds can then not only move the physical matter to which they are wired by neurons, i.e., our own limbs and fingers, but the physical body can build machines and houses. Our limited God-like power to move physical matter remains beyond human scientific proof, so that God continues to hide himself from us. We can believe it or not.

SPIRITUAL TECHNOLOGY

We've come to understand physical technology on mortal earth quite well. Dark matter is eighty percent of the universe. We surmise there are laws and capabilities in spirit matter and resurrected matter more complex that the technology we've discovered inside of a cell, including DNA.

There is a speed limit on photons, which are physical light particles. There is another type of light, a spiritual light, which is, from what religion tells us, obviously not subject to the speed limit of earthly speed of light, any more than the communication technology of prayer is. The ark of the covenant as a communication device would be singularly ineffective if limited to the speed of human light.[44] There isn't much around our solar system within a handful of light years. Zacharias saw the angel Gabriel: *I am Gabriel, that stand in the presence of God; and am sent to speak unto thee, and to shew thee these glad tidings.*"[45] Gabriel did not travel quite like we do.

GOD'S TRAINING PROGRAM FOR MORTALS ON PLANET EARTH

We have mortals running this planet. That's the bad news.

The mortals gain experience. That's the good news.

The system is foolproof, because of the atonement of Christ. That's the best news.

We are mortals for eighty years, prior to being resurrected for fourteen billion plus. God started us out as mortals because the integration of our phenomenal minds with the astounding biotechnology of the human body begins a symphony worthy of eternity. Eternity happens when the spirit's technology is integrated into resurrected, physical, permanent, indestructible, eternal technology.

As mortals, God can allow us to do almost anything. It is the great field trip of eternity, right at the start. He has to clean up our messes and mistakes afterwards, but it is all worth it. We will remember these experiences for eternity—because of the tremendous virtuosity and memory capacity of our minds.

Governments are optional. For some reason it is really hard for seven billion people to be on a planet together without forming governments, rules, and other forms of cooperation.

The president of the United States is the leader of hundreds of millions of people. Kings, presidents, and dictators around the world lead seven billion. We are being trained in running countries and nations. There is enormous room for error. We have endured six thousand years, based on Bible measures, in spite of bloody tyrants and dictators, and in spite of slavery, poverty, plagues, and famine.

Frail and fallible mortals run this planet. We know that beyond a doubt. God gives us the opportunity. He lets us become kings and rulers, congressmen and citizens. We wage wars; he doesn't stop us.

Our modern nations unite to fight genocide and terrorism. We give place to refugees, and food to the hungry.

We aren't just being trained in how to become better individuals, we are gaining experience in running the world.

The eighty years we have as mortals is the field trip of an

eternity. The reason God has us do it before we get resurrected is that after we are resurrected we aren't allowed to sin anymore. If we want to keep sinning, we'll end up with Satan and his followers for eternity. Everyone else agrees to stop being hellions and follow eternal rules.

WE WANT TO BECOME

We don't all become president of the United States, and what could be more important than that? Becoming mothers and fathers is more important. Being part of families is more important.

There is an endless range of things we can become in mortal life. We become pilots and astronauts. We become engineers and scientists. We become best friends forever. We become good neighbors. We become good singers, dancers, football players, and ice skaters. We use our talents. We want experience. We need freedom to have experience.

We are willing to risk going to hell for freedom.

THE GARDEN OF EDEN

God put Adam and Eve in the Garden of Eden so they could choose to take the fruit of the tree of knowledge of good and evil. He knew it was worth it for them to go through mortality with evil, but it had to be their choice; God could not be responsible for inflicting evil on them without their consent. Doing so would be contrary to his nature as a perfect being.

God is good, and he cannot intentionally inflict evil on us without our consent. The only possible solution to the problem of evil is that God provided every single one of us, at some point in our creation, with the opportunity to make the same choice Adam and Eve made. We choose to be mortals in a world full of good and evil. We partake of the fruit of the tree of knowledge of good and evil, the tree of mortality. We decide having evil in a mortal world is worth it.

CHAPTER SIX

HELL

Unless you die when Christ appears at the Second Coming, you won't get to go straight to hell.

RELIGION AND THE AFTERLIFE

I have spent a lifetime disappointed about how little religion tells us about the afterlife. I believe God gave us so little information because he wanted us to focus on mortal life. Sometimes I don't care what God wants: What I want is to know everything, and I don't want to have to die to find out. Maybe at some point he'll share more.

God says, *Ask, and it shall be given you; seek, and ye shall find; knock, and it shall be opened unto you.*[46] I've asked. I've searched, and studied everything I could find, both in religion and elsewhere. I've spent decades asking, trying to peek under the door, pounding, and trying to kick the door down. It hasn't worked. On the other hand, I've learned to organize and clarify the information we have available.

DESCRIBING PARADISE

And he said unto Jesus, Lord, remember me when thou comest into thy kingdom. And Jesus said unto him, Verily I say unto thee, Today shalt thou be with me in paradise.[47]

Jesus wasn't in paradise for very long. After three days, he was resurrected.[48] Paradise in that sense refers to the spirit world; the same term was also used to refer to the final destination of the righteous after resurrection.[49]

If you ask ten people (I've asked a lot more than that) what they think the afterlife is like, they'll all give you a vague answer. What we believe about the afterlife comes from the Bible and from near-death experiences. I believe the two sources are highly consistent.

Our vague idea of paradise is that it is a white place. It may or may not have a garden. It may or may not have white buildings of amazing architecture. The sky is sort of white. There might be clouds. The clouds are kind of low though, so you can sit on them. More like fog, really.

Our vague idea of hell is that it is dark. There is fog, only the fog looks like dark mists. Fog tends to look like dark mists in the dark. There are no buildings, and no sky. We think it's just sort of like being in a dark room with dark mists, but not really a room. You feel like you are in a large room, but you can never find the walls, and you get permanently lost trying; more like a dark labyrinth or maze, but without any walls. And there might be evil beings there in the darkness to terrify you or make you feel guilty, but you won't be able to see them. You might be able to hear them, like evil whispers, or they might put thoughts in your mind to drive you mad.

Our popular culture says Satan and his followers look like satyrs with red fur and red skin, pointed tails, horns, and tridents. They dwell underground in caves with bubbling pools of molten lava/brimstone. This is the alternative to the vague darkness theory: when we die, we will be in hell with Satan in a vast underground cave. I don't think very many people literally believe in such a place; we view it as nonsense. Most people tend to believe in the dark mist place instead. I don't think either one is accurate, or even close to what happens to people on the hell list who die before the Second Coming.

Churches have their own versions of heaven and hell.[50]

THE GREAT SCHISM AND THE REALITY OF THE AFTERLIFE

The most striking thing about the afterlife is that it is reality, not a dream. The great schism between what we imagine and what things are really like is pervasive for mortals, and it persists after we die.

We will find that our existence after we die is as much bounded by reality as our mortal lives are. There are four types of matter: mortal, spiritual, transfigured, and resurrected. The nature of spirit bodies relates to the characteristics of spirit matter, the matter they are made of.

All of our fantasy notions of heaven are hell are like our dreams. Arriving in the afterlife morphs our dreams of heaven and hell into reality.

SATAN IS ON EARTH RIGHT NOW

Satan was cast out of heaven into the earth, with his followers, a third of the host of heaven: *And his tail drew the third part of the stars of heaven, and did cast them to the earth… And the great dragon was cast out, that old serpent, called the Devil, and Satan, which deceiveth the whole world: he was cast out into the earth, and his angels were cast out with him.*[51]

Satan and his followers are spirits; they are "*the angels which kept not their first estate*".[52]

We believe from religion, then, that Satan and his followers look like angels, i.e., people in white robes. Satan didn't change what he looked like simply by getting cast out into the earth. What changed was his disposition, from good to evil.

We also know that Satan and his followers are on earth. Take a look at the earth as seen from space in NASA photos.

You don't see the 7 billion mortals. You also don't see Satan and his followers, but they are on earth just the same.

HELL TODAY

Unless you die during the Second Coming, you won't get to go straight to hell. This is why: Hell today looks like the planet earth. Hell two thousand years ago, at the time of Jesus, looked like the planet earth. Satan and his followers have been on this earth since the beginning. Satan was here when Adam and Eve were.[53] Satan was there with Cain for the first successful murder temptation.[54]

Satan is not in a cave underground. He has been on earth, among mortals but unseen by them, since the earth was in its pristine, fabulous, original state, unpolluted by mortals.

Hell is characterized by the everlasting punishment of being cut off from the light of God.

God views jail differently than mortals do. We put people in jail to get them to stop harming us, or as a punishment. God withdraws his light and love, and the result is hell. It is spiritual death. It is indeed endless and eternal, if you are Satan, one of his followers, or a son of perdition.

Going to hell means being cast out from God's light and love for a thousand years, if you are on the hell list.[55] As of right now, mortals who have died aren't there yet, and they won't be until the Second Coming.

WE ARE SPIRITS BOUND IN MORTAL BODIES; DEATH IS THE SEPARATION

Death is the separation of our spirit from our mortal body: *For as the body without the spirit is dead…*"[56] Our spirits leave our bodies, but our spirits remain on earth.

The characteristics of our spirits, which are present within our bodies, are masked by our bodies.

MORTALS ARE TEMPTED

We know precisely where Satan and his followers are right now. They are among mortals, watching us. They are following us around putting thoughts in our minds. We call that temptation; it is the one power Satan and his followers have with mortals.[57] It is a characteristic of a spirit. It is as much a technology as your ability to hear someone speak; it Is biotechnology, meaning it exists in a living thing, as opposed to a machine or device. Communication between spirits reaches beyond our current understanding of technology. It allows spirit beings, not made of atomic matter, to communicate with mortal beings.

Temptation is a spiritual experience. It is one spirit putting thoughts in the mind of another spirit which is in a mortal body. Satan, who is technologically a spirit, or Type A matter, has the ability to communicate with the spirit or Type A matter within mortals. It is a spirit to spirit communication technology. It is effective in spite of the spirit being inside, or integrated within, a human body.

We all feel temptation virtually every day of our lives, and we are very familiar with how it works. Satan and his followers would not be able to tempt us effectively without being able to see us.

Satan and his followers can influence us as mortals, which is why we know where they are, but we are not in Satan's presence. We don't see him.

AT DEATH, WE ARE IN THE PRESENCE OF SPIRITS

When we die, we will be spirits, just as Satan and his followers are. Some have guessed that there are between forty and eighty billion people who have lived or will live on earth as mortals. When we die, we will be able to see other spirits. To each other, we will look like people in white robes. Satan and his followers also look like people in white robes.

To imagine this, look around. If you were suddenly dead, you would see exactly what you were seeing as a mortal, but you would also be able to see other spirits, who look like people in white robes. The question is what are they all doing?

WELCOME TO HELL

When you die, you will not be personally greeted by Satan carrying a big sign that says: "Welcome to hell. I have you now."

Satan is not going to strengthen your faith by admitting: "I am Satan, and I hate God because he cast me out of heaven."

The reason you don't get a personal greeting from Satan is that he is a busy person, filled with self-importance, who doesn't particularly like mortals.

Satan's followers are not going to greet you with a sign either.

The distinguishing characteristic of Satan and his followers is that they are deceptive. Don't expect them to explain hell to you, and don't expect them to explain God to you. More likely they will tell you there is no God, and that evolution is true, and your spirit was produced as a natural part of the evolutionary process, albeit unknown to humans. If you are on the hell list, they can point out you were expecting to be boiled in lava, and it isn't happening, so there must be no God.

Satan and his followers will know everything you did. They will be aware of the sins you committed. They were there, at least some of them, to watch it. They gave you the ideas. If they were inclined to taunt you, they'd have plenty of ammunition. More likely their response will be: "Don't let it bother you. There is no God. It doesn't matter."

The benefit of being a believer in God is that you can be forgiven. That means you won't care what Satan and his followers know. That is part of the peace and rest associated with being in paradise, but it applies equally well even if you are in the presence of Satan's followers. They have no power over you.

Satan and his followers tend to follow mortals around. They are more concerned with tempting mortals than they are with harassing those who have died. They don't do long-distance temptation for mortals. They get up close and personal. With forty billion of them, they can cover the globe. They can double-team. They take shifts. They are organized, but they are not characterized by having obedient and disciplined nature. For mortals who have died, Satan's followers have the ability for personal confrontation, persuasion, deception, and harassment.

THE HELL LIST

There is a specific list of people who go to hell: "*But the fearful, and unbelieving, and the abominable, and murderers, and whoremongers, and sorcerers, and idolaters, and all liars shall have their part in the lake which burneth with fire and brimstone: which is the second death.*"[58] The list includes adulterers.[59] Keep in mind that those who stop being liars, adulterers, etc., can be forgiven by God, and they don't go to hell. Murderers are the exception. They go to hell for a thousand years, and are then delivered.[60]

THERE IS NO PHYSICAL PUNISHMENT

We do not get boiled in a lake of lava after we die, no matter how many times we committed adultery.

Guilt is a spiritual feeling. We have all felt it, regardless of our belief in God.

HELL IS A STATE OF MIND

Hell is a state of mental anguish after mortal death. It is a condition of remorse.

If you "go" to hell after you die, you won't necessarily know you are in hell, at least until after the Second Coming of Christ, when you will be cast out.[61]

What you will know is that you just left your physical body at death, and that your conscience works much better after you die than it did as a mortal.

GOING TO HELL ISN'T WORTH IT

Religion describes hell, symbolically, as a lake of fire.

To say that most people go to heaven is not to diminish the effectiveness of hell.

To say that hell doesn't include physical punishment doesn't mean that it won't cause anguished suffering.

Satan's punishment is not being able, eternally, to have a mortal or resurrected body. He has spent many thousands of years, approximately six thousand recorded since the beginning of the Bible, on the earth among mortals. His punishment has been the mental anguish of knowing he was cast out, and his progression was stopped. The time will come when he and his followers will be cast out of the earth. He will spend an eternity with nothing but his spirit form, and without so much as a planet to live on. He will be powerless.

The punishment for a murderer is the anguish of a bright recollection of how his or her victims, and everyone who knew those victims, were harmed by the act.

THE NATURE OF HELL

Hell is only natural.

If you eat too much as a mortal, you will become obese. God doesn't have to explain that, it just happens. Most of us have been overweight from time to time.

In the same way, if you do the things that cause you to go to hell, the consequences are as natural as getting obese from too much carbonated high fructose corn syrup water, or if you're a little smarter, getting obese from too much chocolate cake.

Commit a murder and you will feel rotten, naturally. You will feel rotten whether you are mortal or not. Be a heroin addict, and you won't be happy. You might overdose and die. Dying isn't going to make you happy, although it will keep you from using heroin. Dying will make you wish you had some way to escape your pain and your mental anguish. There won't be any way to escape, and most of all, there won't be a way to repent. You may pray, but you won't be able to receive an answer, naturally. That is what is meant by spiritual death, the second death. I believe heroin and other drugs have a spiritual effect, which will be evident at death.

If you are on the hell list and you die now, your experience will be different from those who die at the Second Coming. Right now, you will be on earth among mortals, unseen. If you want to repent, you can pray for forgiveness. The problem is you can't go back and feed the hungry or visit the sick, because you missed your chance in mortality. It is harder to conquer a heroin addiction after you die.

If you die at the Second Coming, Christ is going to be on this earth for a thousand years, and he will put you someplace separate, described as the bottomless pit.[62] At that point you will know you are in hell, and you will also be unable to repent. You will get no answer to prayer for a thousand years. My speculation is that this earth will be removed from the solar system, along with the other planets, and those on the hell list, with Satan and all of his followers, will be left here without a planet. They can go live on the sun if they want, which is pretty much a lake of fire and brimstone. As spirits, they would not be consumed by the sun's heat. Their other option is to stay in space, in the area where the earth used to be, and look at the sun and stars. They can talk to their fellow hellions, and throw parties with no beer or pretzels, or even a nice planet to live on.

Satan's problem is that he can't repent. He can pray all he wants, he can rant and rave at God, but he will get no answer. He has mental anguish that he rebelled against God, and he knows he can't fix the problem. His problem, however, is far worse than for mortals who go to hell, because their misery will end. David, a murderer, was not to be left in hell.[63] From his death thousands of years ago, until Judgment Day, he was to suffer for his murder; at Judgment Day, he will be cast into hell. A thousand years after Judgment Day, he will be placed as a resurrected being on a planet like this earth.

HELLISH NEAR-DEATH EXPERIENCES

Very few people have near-death experiences that are dark and negative. Very few near-death experiences describe going to hell. This is a good sign.

The key elements of such dark experiences include confrontation with evil beings, and being in darkness and confusion.

HELL WITH A LIFE'S REVIEW

If you are on the hell list when you die, your life's review has its own twist: a life's review includes feeling the emotions others felt interacting with you. For rapists and murderers, they would have a comprehensive view not only of what their victims felt, but also of the parents and family members of victims. Having a perfectly clear vision of these feelings, and haing a perfect memory, would place the heinous nature of the acts in continuous and endless clarity and depth for the perpetrator.

This would seem natural, and not an obvious punishment from God, if the life's review is a natural characteristic of being a spirit; in other words, if our spirits are natural recording devices that absorb far more information than our conscience minds do. For example, there is some evidence that spirits do not need to speak to communicate. They read each other minds, and know each other's thoughts and emotions. These emotions could be perfectly recorded, and could be automatically released into the consciousness of every spirit at the time they die. Assuming there is a connection between every person you strongly interact with, and a connection from that person to the others he or she strongly interacts with, would explain the sense of knowing everything at once. Near-death experiences often include this sense of having all the data, including emotions of others, and of viewing it from a more objective perspective. There is a sense of understanding everything at once, of not being constrained by mortal comprehension.

HELL IN THE FUTURE

Hell today exists as guilty feelings and remorse in the minds of those who have died on the hell list. It exists in the mind of Satan and his followers, in their remorse that they were cast out of God's presence. Even though they dwell on the earth among us now, seeing us but unseen by us, the day will come when they will be cast out again.

Hell in the future is defined as a place. Hell will exist as a place for a thousand years. At the Second Coming of Christ, Satan and his followers, and everyone on the hell list, will be cast into hell, a place separate from everyone else, living or dead.[64] After a thousand years, Satan and his followers will be loosed again for a little season. Then they will be eternally cast out, with the sons of perdition.[65] Hell will exist as a place eternally.

CHAPTER SEVEN

WHY WE ARE HERE

MORTALITY BEGINS ETERNITY

Of all the ways God could start our eternity, why did He choose mortality? Why did he ever let Adam and Eve go near a tree that would bring evil into the world?

We are mortals to experience good and evil.

After we die, we will continue to experience good and evil, particularly if we go to hell.

After we are resurrected, we will spend the next fourteen billion years living in peace, and then continue for eternity. God knew mortality was the best way to start eternity, and it wasn't to get the worst part over with.

ORDINARY LIFE

They say when you die, your life passes before your eyes.

I grew up in one of the strongest families in America. My parents were two of the best people I've ever known.

In Cody, Wyoming, mine was one of at least six families with seven or more children which helped shape our community. These families are bonded to each other even now, with a multitude of connections which have grown through the years. The connections include many other good families and good people.

We all have those types of connections. They aren't just part of life, they will continue even beyond.

Cody was rural America in the 60's and 70's. There were about five thousand people, and my family lived ten miles from town on seven acres. When I wanted to go fishing, I would either ride my bike to the creek a mile away to catch small Brook Trout ("Brookies"), or walk a mile through rough terrain to get to the Shoshone River to catch Brown Trout. Rising directly from the river were the McCullough Peaks. On the other side of my house, we looked out at Heart Mountain. When we wanted to hike, we would climb Heart Mountain. It wasn't Heart Mountain because it made your heart race when you were hiking, it was Heart Mountain because it was in our hearts, at the center of our being.

My grandfather was an early settler in Ralston, Wyoming, and he kept sheep on Heart Mountain. In the summer, he would take them to the Beartooth Mountains in Montana. My grandmother was an immigrant in 1912 from Liverpool, England, with soft red hair. My grandparents lived on Heart Mountain when they were young and in love, and they raised their six children there.

My mother spent her life serving, helping, and lifting others. She was a giver. She was a teacher, and there are dozens or hundreds of children and young people, now adults, who will remember her kindness and her love for them. My mother was my hero. I've always hoped I could be like her.

WE ARE GOD'S CHILDREN

The vast majority of Americans believe in God. We agree that God created us as his children. *The Spirit itself beareth witness with our spirit, that we are the children of God.*[66]

THE GREAT SCHISM BETWEEN IMAGINATION AND REALITY IN MORTAL LIFE

There is a great schism between our imaginations and

reality in every aspect of mortal life. For example, as a young man or young woman, we imagine how wonderful it will be to meet the person of our dreams. We imagine falling in love, having children, and being endlessly happy.

The reality is that we do fall in love, we do have children, and we have some happiness. We also have challenges. Marriage is hard. Our spouses disagree with us, often and strongly. The reality is wonderful, in a way. It is wonderful to look at your spouse after 25 years and realize you love him or her more than ever, even knowing everything was far more difficult than either of you ever imagined when you were young and infatuated.

Every experience in life is like that. Why do rock stars and movie stars sometimes become drug addicts? Being stars and having lots of money is their dream come true, but it isn't what they imagined, it is instead reality. In some ways it is better, but it is also hard. They experience the pain involved, the unavoidable challenges of one flawed human interacting with many other flawed humans over money, power, and fame. They experience ongoing pressure and stress. The same is true of professional sports, and so many of life's endeavors. Our dreams don't come true, instead they morph into reality.

That doesn't just happen for mortals. It also happens after we die.

THE PROBLEM OF EVIL

The problem of life is the problem of evil. God is perfectly good, yet God created nature to be full of suffering. Our lives, society, and world are designed with better-than-computer precision to ensure ongoing tension and conflict in us as individuals, and among us as a society, as we work, play, politic, and worship together. Mortality echoes the growth of something enduring and priceless: an individual entity, possessed of moral sentience, certain to live into eternity.

EXPERIENCE IS POWERFUL

When the first man climbed Mount Everest, people wanted to hear his story.

When the first man walked on the moon, people wanted to hear his story.

Mary was the woman who gave birth to the son of God—if we had a chance to hear her story, directly from her, we would want to hear it.

Margaret Thatcher was the Prime Minister of England. People wanted to hear her story.

We recognize the power of the experiences of these people. If we meet a high school girl who is a cancer survivor, we want to hear her story. If a young man cuts off his arm with a pocketknife to save his life, we listen.[67] We recognize the power and value of experience.

LIFE'S EXPERIENCE

We should be really glad to hear that experience is powerful. None of us have managed to get through life without plenty of experience. It is as if God had it in for us. He set us up for the field trip of eternity, and threw in an ample measure of boot camp torture. There is this wonderful mix of joyful experience, like being a mom or dad of a one-year-old child, with the daily challenges of being overwhelmed or short on cash. Most of us have had times in life when we were stretched to our utmost limits, and sometimes we think God did it to us on purpose.

Growing up among strong American families in Cody, I didn't know that I would become a divorce lawyer.

I knew I needed to go to college. My parents taught me that. My grandparents didn't go to college, but my parents did. I wanted to get as much education as I could.

I got a degree in business management, and then went to law school.

I started my career as a prosecutor. Then I joined a law firm and litigated cases, including divorce.

For more than 25 years as an attorney I've dealt with the most difficult challenges in life. My clients have been through one divorce; I've been through hundreds. I've handled many types of cases, including wrongful death, medical malpractice, and civil lawsuits. My criminal cases have included murder, shootings, child molestation, and child pornography. I've studied our legal system from a policy standpoint: Why do we have the laws we have, and are our laws effective?

Knowledge, training, and experience make a difference in life. I wanted to be like my mother. I wanted to lift, help, and serve others, and education helped me do that.

Whatever our training or experience in life, we can all lift, help, and serve others. To me, being a giver makes all the difference, not just in life, but after life.

Being a giver is part of being an American. It is part of being free. It is part of heaven.

HAVING EVIL IS WORTH IT

A mortal world with evil is superior to a world without evil because it allows us to have experience.

Experiencing good and evil is the beginning of our eternity; it is tremendously valuable, and the knowledge will stay with us through eternity.

Stories exist and novels are written because of the tension between good and evil, and mortality is the ideal circumstance for that tension. We are free because we can choose good and evil. Without experiencing evil as mortals, we would never have been truly free to choose good over evil.

We call a baby a bundle of joy. We call an adult a bundle of everything else.

We are minds. Emotional virtuosity is our nature. The

heights of ecstasy and the depths of depression are our common experience. We feel millions of nuances of emotion in our lives. All of it is our reaction to our surroundings. We anticipate. We are embarrassed. We are giddy. We are bored. Our minds are distinguished from those of animals because of our capacity to make moral judgments. Dogs feel happy and sad, and they feel anxiety. They are afraid of thunder and lightning. They don't rob banks. Without evil, our capacity to make moral judgments would be wasted. God considers evil to be valuable. Subjecting us to it for eighty years of mortality is so valuable he uses it as the beginning of our eternity. Since he can't inflict evil on us, we had to make the choice.

Emotional virtuosity is wasted without evil.

This world is designed as the ultimate mortal experience. Experiencing good and evil will inevitably teach us what we want to be throughout eternity. Not surprisingly, we don't all want the same thing.

TO BECOME OR NOT TO BECOME

The purpose and end of mortal experience is to become.

Here is a story of a lady we'll call Eva. She didn't get a medical degree or a law degree. I don't think she got a degree at all. Over several decades of life, she married, had a son, and got a divorce. She remarried to a kind man, who worked hard. They started a business. They got sued over the business. She felt threatened by financial destruction. She felt despair. She settled the suit on terms she could not afford, and shouldered the horrible burden.

There is a certain type of person who in a lawsuit will not negotiate or be reasonable. Eva was faced with such an adversary. In her despair, Eva desired for her adversary to be punished by God, preferably in hell.

Finally, Eva called someone and told her she would just

kill herself. That person called the police, and they went to see Eva. Eva truthfully explained to the police that her suicide threat was not in earnest.

The heart of this story is why Eva had the strength to go on, in the face of such despair. She wished she could die, thinking it would be easier, but she never seriously considered it.

Her son was the reason why. Eva's remarkable strength and resilience were developed over several decades of caring for her severely disabled son. She had become too strong and too powerful to consider taking her own life, even in her moment of the greatest despair of her life. We all have that moment, although we don't necessarily know which one it was. Sometimes we have a dozen to choose from.

I learned from Eva. I looked at her and saw the person inside.

For every father who has ever had a daughter, there is a brilliant light and innocence in a little girl. There is no fat or thin, no smart or dumb, there is just the shining soul of the little girl, embodied in in a child's appearance with locks of hair, Easter dresses, and the word daddy.

I could see in Eva an innocent girl was still there, but she had become more. Her experience had given her strength and wisdom. The innocence was dim. The pain was awful. The power was Christ-like. The power was experience. I've seen this same thing happen, over and over again, as the women I grew up with, and the women I see every day, young and old, become powerful, through experience. They become spiritually powerful, and eternally powerful.

CARING FOR OTHERS

I heard of an elderly man who spent twenty years caring for his invalid wife. They will eventually be together when they die. After the resurrection, they will be in the presence of Jesus Christ. Is it easier to be the caregiver, or the invalid? Both have become powerful at loving, caring, being grateful, and suffering. He suffered for her and with her. She was grateful for him and with him. Which of them gave more? They both gave their all. They passed through a refiner's fire.[68] They were as refined as mortals can become, and I don't just mean they had good manners. They were spiritually refined, through experience. They are as pure gold, with no dross left. We live for eternity after our mortal life. Theirs will be a better eternity because of what they experienced, and because of what they gave. They have learned gratitude for twenty years of invalidity. They have become valid.

THE BIG TRIALS IN YOUR LIFE

Every person I know has some big trials in his or her life, trials that go beyond the worries about their children having the flu or not doing their homework and failing a class.

It may be the death of a spouse, or a child. It may leukemia, or a heart defect. It may be years of marital difficulty, followed by divorce. It may be failure of the family business, with bankruptcy, or prolonged unemployment.

These trials are problems we cannot control. It is beyond our mortal ability to avoid them or fix them, so we just endure them. It is God's way of showing us all that we depend on him, not just for the air we breathe and the planet we live on, but for the major events in our lives that are beyond our control.

Like Eva, the big trials in our lives make us become better people. We may, through long suffering, become really good grandparents. We may become really good people. We become wiser, more caring, more loving, and more understanding.

We often hear of people who have experienced a life's

review as part of a near-death experience. In our own life's review, we will see how the big trials in our lives, the ones that lasted a decade, or nearly killed us or a loved one, made us better people. We will see the value of them.

Do we truly believe we are going to live forever? I think we do. We are told we will all be resurrected. We are all supposed to recognize who Christ is, and be redeemed by him, at the Judgment Day. We are supposed to spend the rest of eternity living in the utopia we have desired but failed to find in mortality. God starts out our gift of eternity by having us go through some really hard things as mortals. We all become experienced veterans of life. It is inevitable. The system is foolproof. The hard things that happen, continuously, in our lives, don't just prepare us for harder things in our mortal future; they prepare us for eternity. God never intended for it to be an eternity of punishment. It isn't about religion. It is about a being who is sharing with us everything we are willing to accept, on our own terms. Religion means accepting him, but it is not mandatory for happiness. He will continue to share for eternity. He intends to help us all become very happy, eventually, for eternity. Every knee will bow, and every tongue will confess that God is God, not because God is a dictator, but because he will show us what he has done for us. He will show us the truth of who he is, and that it isn't about him. It is about us. He has invited us all to become his children, and to come to him. He has invited us to come home. He is preparing us to do that. If we choose not to have a relationship with him, he will accept that. He will still forgive us, resurrect us, and give us a planet to live on for the eternities. He has perfected utopia, and will give it to us all, with or without him. If we reject his religion, which is an offer to have a relationship with us, we will at least understand his technology, and we will know he has been fair to us.

WE DON'T ASK FOR BIG TRIALS AS MORTALS

It is really rare for someone to pray that they'll have a big trial in their life. It seems like an unimaginably stupid thing to pray for: "I need to be a better person. Let me have cancer," or "I've decided experience is tremendously valuable, so I need a big trial in my life. I pray my spouse will divorce me."

The big trials happen anyway. They are programmed into our lives with computer precision.

THE GOOD NEWS

The reality of our personal lives becomes clear. It is so different from our childish dreams and imagination. Life is so hard. Life is a constant challenge for all of us. The smartest among us are stretched to their limits; the less smart are also stretched to their limits. The stretching process is tailored to our abilities and capacities.

The majority of people in the world have been able to live as good people.

In mortality, things are hard. Poverty has abounded in so many societies for so many centuries. But those who are not thieves and liars live out their lives as fathers and mothers, as friends and family members, and they gain many mortal experiences. When they die, they no longer hunger or thirst. They find a measure of peace.

Granted, many mortals are overcome. They lead miserable lives of drug addiction, prostitution, theft, or violence, and they die in the process. Some of them live long enough to change and become better people; i.e., they repent. Others don't.

When people who are on the hell list die, they will know the remorse that is hell. They will have a measure of peace, however. They will no longer be subjected to the violence or horrors or hunger that they endured in mortality.

For all of us, we learn that being a wonderful grandma is more important than being a movie star. Being a dad is more important than being a billionaire. We learn that the real value in our lives is measured by our relationships and our families. Becoming a loving, giving, serving, mature person is more important than titles, wealth, or positions.

CHAPTER EIGHT

THE AMERICAN DREAM

THE AMERICAN DREAM

The American Dream is the dream of heaven on earth. All we ever asked for was utopia.

THE AMERICAN DREAM AND AMERICAN REALITY

There is a great schism between imagination and reality in America, just as there is a great schism between imagination and reality in our mortal lives, and just as there will be a great schism between imagination and reality after we die.

The reality is that we have the greatest cities and the greatest wealth in the history of the world, and not just in America, but throughout the earth. Our forebears envisioned these gleaming alabaster cities,[69] but they did not know about the wars that we have had. They thought there would be challenges to freedom, but they couldn't have dreamed of the apocalypses we have known. Our realities don't just include the good, they include the evil, and the dark. They include the decay and filth in our cities, and the crime. Our realities include nuclear bombs, and names like Hiroshima and Nagasaki.

We remember the dark hours of the revolutionary war; our founding fathers and mothers lived through that reality. We have gone on, generation after generation of Americans, to live through the Civil War, the Great War, and World War II. We have seen blood and destruction beyond what our forebears could have imagined.

In reality, we value the great things America is, amidst the great evils we have witnessed. Our experience is reality; our dreams have morphed and there is a bigger reality we fail to see. But if we had not imagined the great future of America, we could never have lived it.

We stand as the divinely privileged witnesses of the greatest society mankind is capable of, in all its splendor of good and evil.

AMERICA IS HEAVEN

If we could see for one moment the vision of what America truly is, we would glimpse that in this one instance, the truth and reality of America is beyond what we have imagined. America is what God imagined, and his imagination is better than ours.

When it comes to countries, America is heaven. No other nation has ever enjoyed such prosperity. Measured by our gross national product, or our product per capita, or whatever you want to measure, we have had wealth beyond ancient Egypt or Babylon. Far more important even than physical prosperity has been the spiritual power of America.

THE GREAT IMAGE SEEN BY DANIEL

Interpreting King Nebuchadnezzar's vision, Daniel described a great statue: *Thou, O king, sawest, and behold a great image... This image's head was of fine gold, his breast and his arms of silver, his belly and his thighs of brass. His legs of iron, his feet part of iron and part of clay.*[70] The head was the gold of Babylon. The feet were miry clay. Is America the miry clay? When I was a child, I thought America was part of the iron and miry clay of the feet, along with the nations of Europe, some strong and some weak. Even as a child, I was offended for America's sake. This didn't sound right. We have had greater wealth and strength in America than any country in the history of the world. There has never been an army stronger than our nuclear missiles. Most importantly, there has never been a country of greater spiritual power, with the exception of ancient Israel.

The great idol represents the strength of earthly kingdoms; some have linked the kingdoms historically to Babylon, Medo-Persia, Greece, and Rome,[71] with the miry clay being European nations after the demise of Rome.

THE STONE CUT WITHOUT HANDS

Daniel described a stone: *Thou sawest till that a stone was cut without hands, which smote the image upon his feet that were of iron and clay, and brake them into pieces...and the stone that smote the image became a great mountain, and filled the whole earth.*[72]

The stone is the kingdom of God. Some might ask whether it was set up when Christ was on earth, and whether striking the image occurs at the Second Coming.

THE SPIRITUAL POWER OF AMERICA

America is not part of the iron or miry clay in Daniel's vision.

America is part of the stone that smashes the idol. America is the great precursor nation to the kingdom of God. Christ owns America. It is the embodiment of freedom, and of all the principles of governance he has given us. This is his country. He helped us found it. It will not fail. If our enemies bring us down, he will come to our aid, and he will establish America forever. America is the country of Jesus Christ. It is the kingdom of Christ on earth. He has allowed us to choose our leaders, and we might have made some mistakes.

To understand why America is the stone, we look at history. I do not mean that America is the stone and that churches are not. I mean that every person who has sacrificed to move the work of God forward is part of the stone. We are all part of the stone, and we are God's. The founding of America, and the establishment of freedoms, including religious freedom, were part of what started the stone rolling. The Second Coming is the point where that stone smashes the idol, and the stone becomes worldwide governance of a transfigured utopian planet. We've been told of the technology that will prevail: medically, there will be no more illness; agriculturally, there will be no more hunger; politically, there will be peace, freedom and unity; and religiously, there will be freedom. This won't because we make scientific advances. It will be because God will share these technologies with everyone on earth, in a single day. He has shown us before that these technologies are readily available. In both the Old and New Testaments, the sick were healed and the dead raised. We can't complain to God that he didn't tell us he had the technology. We just weren't listening.

A STONE CUT WITHOUT HANDS

The word apocalypse means the revelation of God. That is, God will be revealed. He will show himself to us.

God is to show us, in a single day, that he has the power of God. He will show us that he has the technology to create earths, and to control life. Presumably, he will explain whether he took billions of years, or seven days, to create the earth. Either way, he is a scientist who has perfected all technology, and become omnipotent. He has the technology of God.

God founded America by inspiring its leaders.

HOW THE STONE GREW

Jesus Christ had relatively few followers during his lifetime. When he died, the apostles carried the work forward. John lasted perhaps the longest. Between 70 A.D. and the council of Nicea, early Christians had to keep belief in Christ alive. We owe them a debt. They were sincere. They were willing to be persecuted and die.

The reason we have a Bible today is because the Catholic Church preserved it through the centuries. We owe them a debt.

God used anyone and everyone for his purposes as the stone grew. For example, Henry VIII, King of England, was not a savory character. He was on the hell list. He killed Ann Bolyn for the sake of trying to marry someone else. Yet he furthered the stone by establishing the Church of England.

The Protestant Reformation was part of the stone.

Some gave their lives to help the Bible become more available, notably William Tyndale, killed in 1536.

From a religious and biblical standpoint only, it has taken thousands of years for the name of Christ to become known as it is today.

It is a wonderful thing that so many people in the world's religions believe in, and worship, God. The idolatry of the ancients, described in Biblical history, does not prevail in the world today. Christians, Jews, Muslims, and Hindus have strong faith in God. Buddhism has wisdom and spirituality.

The stone was cut without hands; God furthered his work by every available means.

THE STONE INCLUDES ART, CULTURE, AND SCIENCE

It would be a mistake to interpret the stone merely as religion.

What kind of kingdom would God have, if it were full of uncultured ignoramuses?

God sent forth a deluge of information, art, culture, and science, so that his kingdom would be set up with knowledge and power. He did not leave the world in the dark ages. He is bringing us all to the light, and has been for centuries. He just had to avoid doing it too soon.

TECHNOLOGY CONTROLS THE PROGRESSION OF HISTORY

Timing is everything when it comes to technology.

It wasn't just advances in Western thought, religion, art, and music that paved the way for the birth and growth of America, it was also technology.

The earth is a master-planned community controlled by the timely provision of technology. America is the earth's ultimate society built by mortals for mortals.

The technology is released through inspiration, the provision of advanced ideas.

There was a build-up of information and technology, a progression of freedom and true principles, which paved the way for the greatest society on earth, as found in America and her allies.

The technology for great art and sculpture has been present for hundreds of years. Great masterpieces required oil paints and sculpture techniques developed hundreds of years ago.

Our finest music is in symphonies written by Mozart and Beethoven, not in synthesizer music or electric guitars. We use the violins, horns, and the drums developed hundreds of years ago. We have added to the great classical music a world full of music and instruments.

We had the industrial revolution, and modern engines and agriculture.

We have seen the progression of transportation from horse and wagon to space shuttles.

Information has gone from ancient scrolls of papyrus to supercomputers.

All of these advances are part of a carefully orchestrated timing of technology, including weapons.

The timing had to be carefully orchestrated because of weapons.

It is no accident that nuclear weapons were invented for use in 1945 and not in 32 A.D. After less than a century, we are on the verge of losing control of nuclear weapons; we are in danger of nukes being obtained by terrorists.

For nearly six thousand years of recorded history, the world was based in agrarian society, up until the past several centuries. Then, over the course of the hundreds of years leading up to and following 1776, the birth of America, there has been an outpouring in every field of human endeavor.

As a result, America has become the shining beacon of what mankind can achieve on earth. It has become the ultimate society. It has been joined by many allies, most notably Great Britain and France, allies in great wars. America has been the spiritual homeland for the free and the brave from every nation on earth.

Jesus Christ loves America. He loves this country more than we do, and that says a lot.

This is Christ's kind of place. He gave us abundant resources. He gave us strong families. He taught our founding fathers the best principles on earth by which to govern a country, including a constitution and a republican government. Freedom. Choice. Religion. Accountability. Integrity. Struggle. Sacrifice. Reward. The reward is the greatest country on earth.

THE BIRTH OF AMERICA

We the People of the United States of America, in Order to... secure the Blessings of Liberty to ourselves and our posterity, do ordain and establish this Constitution for the United States of America.[73]

As stated in our Constitution, we sought from the very beginning to have God bless America because we wanted to be free, and we wanted to have strong families—which is why we say up front we want liberty for ourselves and our posterity, i.e., our families.

Freedom and family have been our highest priorities since our nation was founded.

GOVERNMENT BY THE PEOPLE

America is governed by its people. This is the great principle and idea of freedom.

Kings and tyranny ruled for six thousand years of recorded history, with notable exceptions.

We took the ideas of Socrates, Aristotle, and Plato. American founding fathers like Thomas Jefferson, John Adams, and Benjamin Franklin developed the ideas they had learned from the generation before them, including John Locke, to put together a practical constitution to govern a modern republic.

The British Empire was part of the stone seen by Daniel. The Empire spread technology and economic growth throughout much of the world. The Empire wasn't perfect, but It served its purpose. God used it to bring about good. When it came time for the birth of America, the colonists revolted from the Empire, with God's assistance.

NATURAL RESOURCES

It is no wonder that America enjoyed astounding growth

in prosperity for hundreds of years. In addition to the abundant advances in science, art, music, medicine, literature, technology, and everything else, the United States was a country rich in resources. America developed those resources, overflowing and overwhelming Native Americans. The injustices by Americans to Native Americans were the bad news, but there was also good news. Our advancements in society have been accomplished in a mix of both good and evil, like almost everything mortals do.

Not only did America have its own wealth of resources, America soon had access to all the resources in the world through the great ocean vessels that technology brought. America was benefited by the fact that it was separated from the countries of Europe and Asia; it remained isolated until 1492.

THE RULE OF LAW

The possession of principles, ideas, and truths is the reason America is the greatest nation the world has ever known. Foremost is our religion. We didn't do it all ourselves. We are a nation under God. We are God's country. We recognize how our religions have worked toward the preservation and establishment of the truths upon which this country was established. We see our abundance as blessings, not a divine right, but a divine gift. Freedom is a divine right, but we as mortals were incapable of creating a society where freedom could be established without divine gifts, and those gifts included religion, technology, philosophy, art, music, science, and the rule of law.

America exists because of the principle of the rule of law.

Mexico and South America have an abundance of land and resources to match the United States, but they have not accomplished what the United States has. There is a wealth of resources in Asia and Africa, but America is not there. What they lack is the rule of law. Corruption in government is a worldwide epidemic. America fought corruption, and still fights it today. We prosecute those who offer or take bribes. We can and do put them in prison. The rule of law has made an essential difference.

ZENITH OF SOCIETY

All of the blessings of the world are found in America and her allies. We have experienced utopia. We have had the greatest nation in the history of the world.

It has been as if God designed the entire history of the world leading up to this moment, the greatest in history, with all of society at its zenith, and with America standing as the Statue of Liberty.

God founded America as the masterpiece of what fallible mortals can accomplish, given the maximum dose of resources and knowledge which can be made available to humans.

America is the ultimate national experience; it has allowed teeming millions the utmost in a mortal experience, the great field trip for eighty years each at the beginning of eternity.

Our world has great treasures of art and music, wealth untold. We have gold and silver beyond Incan dreams of cities of gold, and diamonds, gems, and minerals. We have jet planes and supercomputers. We have all the knowledge of the planet in our textbooks and our computers, and we understand the workings not just of the earth but of the galaxies throughout the cosmos.

STANDARD OF LIVING

The average American has a higher standard of living

than the kings and queens of 500 years ago. A king would send messengers to nearby countries, with weeks delay. We can talk to someone in China in seconds, halfway across the world, on our cell phones. We have news access worldwide. We live with central heating and air conditioning in our workplaces. Kings would ride horses through mud. We have paved roads, and comparative luxury and comfort in our cars. Kings would assemble orchestras. We take our pick from orchestras on our CD players or on-line. We have modern medicine. Henry VIII could have only wished for the kind of medical care we have, when he got sick and died in 1547 at age 55. We eat bananas and mangos from what were once exotic countries, and fresh fruits and vegetables year-round.

THE VAST WORLD OF SPIRITS

Living Americans aren't the only ones who watch America as the ultimate mortal nation. All Americans who have died, and all others who have died, are also watching what happens to America and to the world. They remained on this planet when they died. It is likely they are among mortals, unseen by us, but able to observe everything about us, including our science and culture.

For everyone who has died, from the days of Noah until now, the outpouring of knowledge among mortals which began with Christ's mortal ministry and has continued to the zenith we see in America and in the world today, has been reflected in the spirit world.

All those who have died have been learning for millennia of Christ, just as mortals have.

Those who have died have likely been watching the expansion of science, art, and culture among mortals. They got to watch nuclear bombs being built, to the extent they were paying attention. They got to watch the mapping of the genome. They have seen all the things we have done and accomplished. The advance of earth's technology has happened among mortals, but those who have died are aware of it all. We have all, the living and the dead, been advancing in knowledge of Christ and technology at the same time.

There was a great gulf among the dead before the resurrection of Christ: *And beside all this, between us and you there is a great gulf fixed.*[74]

Christ bridged that gulf when he was resurrected. At his death, the veil of the temple was rent: *And behold, the veil of the temple was rent in twain from the top to the bottom.*[75] This is a classic dual meaning: it happened in the temple in Jerusalem, but it also happened in the heavenly temple. The earthly temple resembles the heavenly temple, the kingdom of heaven. Christ was first to go through that veil and return to God. The result is that Christians in the spirit world have had the same amount of time to spread the word about Christ as mortals have. *For for this cause was the gospel preached also to them that are dead, that they might be judged according to men in the flesh, but live according to God in the spirit.*[76]

The final curtain is not drawn back, either for mortals or in the spirit world, until the Second Coming of Christ. That is the point at which all will know who he is. I am not saying that no one, living or dead, has seen him since his resurrection. They have.[77] Some have had near-death experiences, and describe seeing him. I don't doubt it. I am saying that it was a personal appearance. Christ is not openly among those in the spirit world, because they must live by faith, just as mortals do, until the Second Coming of Christ.

CHAPTER NINE

AMERICA AT RISK

AMERICA IS GREAT

America is great. America is great because God put the best people and the best leaders in the history of the world here. He spent millennia pouring out information, science, religion, and technology so that we as mortals would have the finest field trip of our eternity. America is the ultimate mortal nation. God helped us set it up, which is why we say it is part of the stone, the Kingdom of God, cut without hands. We could not have done it without him.

America has always been great. It was born great. It is still great today.

AMERICA IS STRUGGLING

What happens when you take America, a country God dearly loves, which was set up by him as the finest example of what humans can accomplish when given every possible advantage, and you confront that great nation with the worst evil the world has ever known?

You get the presidential election of 2016.

The challenges in the election and in our country were staggering. I don't mean that the political candidates were bad; I mean that they were confronted by nearly insurmountable problems. We had congressional gridlock; intractable terrorism in the Middle East; decaying infrastructure; staggering national debt; and arguably the most rancorous and divisive election of all time. Part of the rancor stems from the extremely difficult challenges we face.

If God weren't in charge, I'd be worried. We've already seen what he did to get our nation to where we are now. We've seen the deluge of light, knowledge, and information he's poured out on us. God is aware that you can't allow mortals to stay on earth indefinitely because they will eventually be overwhelmed and fail. Timing is everything.

AMERICA WILL NOT GO TO HELL

The way to keep our country great is to maintain our values and the rule of law.

Countries can go to hell, figuratively. A country goes to hell when it is ruled in tyranny, or when it is so filled with corrupt officials that the rule of law fails. This isn't caused by national debt, so much as it is caused by bankrupt values.

There is a difference between a person who makes a mistake and is struggling, and a person who is going to hell.

There is also a difference between a country which makes mistakes and is struggling, and a country which is going to hell. America has made mistakes.

DIVISIVE ISSUES

Some of our challenges threaten our existence as a nation, and some do not. Current issues in the U.S. include racial tensions, including policing and criminal justice issues. We are concerned about political gridlock, loss of employment, and economic problems. There are immigration issues. We had a recent refugee crisis. We are concerned about terrorism and the global economy.

In the U.S., religious freedom versus personal choice is a clash of titan principles. The issues transcend us as individuals. The divisions among us have included racism, abortion, gay marriage, prayer in public, and creationism versus science. Divisions are part of America, and always have been. I am not suggesting that the problems aren't serious. I believe they are profound and pervasive, and we must do everything possible to address them. These issues are not imminently threatening our existence as a nation. They are ongoing issues we deal with.

In *Roe v. Wade*, the United States Supreme Court did not know precisely under what circumstances it was moral to allow abortion. The justices today do not know. None of the rest of us know. That is precisely the problem which causes the division. We abide by what the justices decide. The alternative would be the downfall of the rule of law, and of freedom. We are aware that we rely on fallible mortals.

Morality is the art of knowing what God would do, i.e., omniscience, and we are faced with situations in which we don't know. For the atheists and agnostics among us, they don't believe God has anything to do with morality. For the rest of us, God hasn't given us the details on where to draw the legal lines. He's chosen to leave the decisions up to us.

We have the power to avoid unnecessary contention. We can and should be united in the love of our country and the principles of freedom and family. We should avoid violence and contention as one of the most important things we can do for our country. *Finally, be ye all of one mind, having compassion one of another....*"[78] As a country, the key to avoiding violence and contention is in the rule of law. We vote on the divisive issues. We maintain free elections. We abide by laws we pass.

AMERICA AT RISK

Will America cease to exist as the greatest free country in the world because we vote to surrender our freedom? It is possible, but I hope not.

America is at risk because our families are at risk.

Having families at risk increases the burden on our government and economy. Broken families correlate to increased incarceration rates, lower education rates, drug use, and increased welfare. Having two married parents in a family, instead of single parents, is statistically unbeatable.

If families were the only concern, we might be optimistic.

We have other challenges. Terrorism is an ongoing, serious threat. Having unstable countries with nuclear weapons is a threat. Washington D.C. is the most likely target in the world for a rogue nuclear attack.

Our freedoms in America are at risk. Our infrastructure is aging.

Our political system is paralyzed. The key cause of the paralysis is our national debt. That is the existential threat to America.

If we see America knocked to her knees, it is because of the tremendous economic burdens and obligations we have strapped upon her. Americans can't begin to afford to pay our national debt. It is increasing, and we feel powerless to stop it. America was born on July 4, 1776, at the signing of the Declaration of Independence. After two hundred forty years, our debt has escaped us, and our system is malfunctioning. There is anger in the country, anger at the establishment, and political gridlock.

We have been the world's leading superpower in weapons and military, but we cannot maintain that position if we fail economically.

Our children's economic outlook has dimmed because we didn't control our debt. We see pervasive discontentment and anger with our political system.

There is an additional reason for our declining economic outlook: demographics have shifted. The key resource now is labor, and labor is cheaper overseas precisely because of our prosperous standard of living. We still have natural resources available; we can buy iron ore, copper, and other commodities, and the costs of those resources is based on global pricing, but labor is not.

Our economic systems are being changed by the increasing connections in the digital age, with a global flow of data.[79]

What comes after a zenith is darkness, and we feel and fear the threat that America will decline into chaos.

We are concerned that America is so burdened that she may be overcome. She may fail. She may be defeated. We may lose her, our beloved homeland. We hear echoes of impending disaster, taunts, and jeers. We have nightmares of fires, terror, and pillars of smoke, and we don't know if it is post-traumatic stress or premonition. There is a certainty that an out-of-control national debt will eventually destroy our government. It is increasingly unlikely that we will get it under control. The question is how long it will take, and whether we are moving in the right direction.

After two hundred forty years, it may well be that our national debt has escaped our control. Our system of government may fail.

ZENITH IN THE GARDEN OF GOOD AND EVIL

We have the greatest cities the world has ever known, and the greatest pollution, violence, and corruption the world has ever known.

America, the great Republic, is the pinnacle of worldly freedom. We couldn't have done it without our allies. They couldn't have done it without us. We are one. We have all been part of the great fight, a great war between freedom and tyranny, between liberty and despotism. We are united in freedom and for freedom.

Our love of freedom and family is what we all have in common. It is our spirituality. Every human feels a love for others. It is the connection between us that identifies us. George Washington was our symbolic prophet as much as our president, in this country God loves. So was Abraham Lincoln. The revelation George Washington and the founding fathers received was a form of government. Abraham Lincoln's revelation was how to save the union, America, and freedom itself. They did as much as any of us ever has in rolling forward the stone. They were inspired by God. They served God. *Now the Lord is that Spirit: and where the Spirit of the Lord is, there is liberty.*[80]

Our soldiers fight and die for our freedom. Our police officers, firefighters, and first responders are prepared to fight and die for us. Americans are good. We are known for doing the right thing. Others fight to establish tyranny. We overthrow tyranny, and give back freedom. Our military has become the most extraordinary fighting force in the history of the world: technology has allowed us to shock and awe, but our military is more than powerful. Our military is extraordinary because we fight not to take over the world, but to free the world. America is a prototype of what freedom should be, and it is the best country that flawed humans can achieve. If America falls, it will be our fault, as Americans. If we are not good enough, wise enough, and moral enough to save her, the blame will be on us.

An American is a person who will give anything and everything for freedom and family. An American is a person who is sacrificing to help roll forward the stone cut without hands, the Kingdom of God on earth.

Not all Americans are from America. I consider any person who lives and dies for freedom an American in spirit, no matter what his or her native country is. We have allies. They live in every country on earth. In every country on earth we can find people who are willing to fight and die for freedom. Many have made the ultimate sacrifice: they have died speaking and acting against tyranny. In most countries those who love freedom are outnumbered and outgunned. Not in America.

More than seeing freedom-lovers in other countries as Americans, I see America as freedom itself. Americans are freedom.

Moral freedom is society perfected.

GOD USES TECHNOLOGY TO PROTECT HIS PEOPLE

God uses technology to protect America.

In the Revolutionary War, he helped us fight an earthly king who oppressed us. He sent us friends like France. He sent us the greatest leaders he had, including George Washington. He sent us patriots. He didn't send us saviors, he was our savior.

In the Civil War, he helped us get rid of the evil of slavery. He didn't do it for us, he helped us. He put the most resources and the best leaders on our side. He sent us Abraham Lincoln. Americans fought for freedom for all.

In World War I, good people from all over the world gathered to fight for freedom, and preserved it.

In World War II, God gave us superior technology. We had radar. We had codes, code-talkers and code-breakers. We had the great generation. We had nuclear weapons first, not by accident. Our allies fought on our side, and we were all God's people. There were those even in Germany who were among God's people, who suffered much through no fault of their own. Christ knew they were on his side, and it was through their willingness to be there and experience that suffering that they served him. We saved freedom, and he was with us.

We fight now against terrorism as the world has never known it. Nuclear weapons did not eliminate evil and tyranny, but they changed the format of war. We fight where we are needed around the world. Our soldiers, the men and women of the military, fight for us. Some make the ultimate sacrifice. Many pay a price of lasting mental and physical suffering. We owe them a debt.

We continue to have the best technology in the world at our disposal.

War is ugly, and our soldiers and our country have made mistakes, but we, the people of America, still stand for freedom, and we lead the way. If not us, no one. America is and has been the leader of free people. A multitude of people and nations have stood with us to fight for freedom, and they will keep standing with us even until the very end. We will never give up. We will stay free, for we are America, and God is with us.

Can God help our country? He can. He has. He will.

Can he help you, individually? He can. He has. He will.

Can he help you through every big trial in life, including the ultimate sacrifice, if necessary? If victims had not agreed to be mortals even knowing there was a risk they might die, God could not have placed them here where they would suffer. Every murder victim dies to save us all, but not from hell. We are saved from inexperience. We are saved from a world that might not have been--this world. We are saved to live in America. We will be saved to live in heaven. Everyone who has died for others, in war or in violence, is a murder victim. They died as givers. They make the ultimate sacrifice. Can God help us all, through everything? He can. He has. He will. God is a giver. He is a sharer.

Will God save us from hell? He will if we ask him.

After about a thousand years, everyone will ask him, except the sons of perdition.[81]

OUR ETERNAL COUNTRY

The great blessing of mortality is being able to make mistakes as a country as well as individuals. We will do our best as mortals to govern our own countries, not just in America, but in Great Britain, France, Germany, Japan, and every country. We are here for the experience.

Eventually God will save America. He will fix our mistakes. He will make this country his own. We will be his people. Those in paradise will join us, because he governs them as well.

We are here as mortals to have the experience of running a country. After we die, this experience will help us understand how God governs planets, galaxies, and the universe.

A CONFESSION FROM MY GENERATION

The presidential election of 2016 was driven by anger. Anger stems from the pain of being hurt.

The rising generation of adults, from age 18 to 39, had been hurt by us, those 50 and older.

We confessed to the rising generation that their future was dimmer than ours. We confessed that our congress is paralyzed and malfunctioning, unable to solve our most serious problems: our national debt, and our national and global economies.

We told these things to our children and our grandchildren, and they were hurt. The profound depth of their hurt stemmed from their sense of the seriousness of the problem. Even if we tried to lie, or refused to acknowledge these problems, the truth was evident in our economy.

There was a backlash against the politicians and the system.

In fairness, the fault does not belong to all Americans age 50 and over. Some of the Americans who led the way to our current crisis died long ago, and many Americans never voted for the policies. It is not a generational issue, it is one of the weaknesses of a free republic. Benjamin Franklin said the founding fathers had given us a republic, if we could keep it. We sense that we risk losing it.

I was talking to a young friend of mine about our economy and the future. He was about 30 years old. He asked me if there was anything we could do to fix the problem of America's debt. I said no, it is a function of the age of our nation and stage of our debt levels. I said it didn't really matter who would be elected president in 2016, in the sense that no one could solve this problem. To clarify, that isn't absolutely true. We can fix this problem as a country. All we have to do is to have a majority of Americans vote to pay more taxes and/or lower spending so we can manage our national debt, without causing a national or global depression.

THE GRAND SOLUTION FOR AMERICA

The more likely solution to our problems as a country will be discussed in the next chapter, titled "The End of the World."

America is not a casual participant in world events. America is the leader and bellwether, and America is in distress.

What if we lost America?

The world would be hell.

CHAPTER TEN

THE END OF THE WORLD

We want to know what comes next.

Individually, we tend to view death as the end of the world. We want to know what happens after we die. What will we experience in the afterlife?

We also want to know what the future holds for America, and for the world.

We want information about the end of the world.

AMERICA'S EXISTENTIAL THREAT

Will America be overthrown? Will we be overrun by foreign invaders? Will we be defeated militarily? Will America cease to exist?

The question of whether America will be invaded or overthrown has to some degree been determined by nuclear weapons, the paradigm shift after Hiroshima and Nagasaki. We haven't been invaded by huge numbers of troops, ships, and tanks, because any country which attempted it would be vulnerable to nuclear destruction of their capital and other major cities. An invasion by a non-national terrorist force could be different. Such a force may not be deterred by nuclear destruction in the home countries of the participating troops.

THE ZENITH OF SOCIETY

We are at the zenith of society on planet earth.

We know this because Moore's law failed in 2016.

Moore's law was a guess by a guy named Moore that we would double the number of transistors we can effectively use in computing devices approximately every two years. This actually happened for decades. We can now fit billions of transistors on a microchip.

I am only partly jesting in saying there is a relationship between Moore's law and the zenith of society.

God gives us technology on his timetable.

He gave us nukes, and we started nuclear warfare in 1945.

He gave us the information to map the genome in 2003.

He gave us the Large Hadron Collider to discover the Higgs particle in 2012.

He gave us microchips which maxed out their capacity in 2016.

The information explosion has had a profound impact on our society. Even if we develop anti-gravity cars to fly in, the impact will be less. If we alter DNA to eliminate aging, we can all live to be a thousand years old. That would have quite an impact. I'm not sure it would be a good thing.

THE INFORMATION EXPLOSION

Look around and see a society at its zenith. We have an overload of data and information. We can watch videos on our smart phones. We can look up information and answers to any and every question. It isn't making us a more peaceful society, but it is certainly exciting to be a part of it.

THE END OF THE WORLD COMES AFTER THE ZENITH

The end of the world comes after the zenith; the question is how much time we have. Let's all hope it is more than a few decades.

ATHEISTS

For those who don't believe in God, there is no evidence that the end of the world is coming within a century or two. It is enough to say that God is a scientist, and that he designed mortality as a technological paradigm.

For most world religions, the end of the world isn't an issue. For Christianity, it is paramount.

RELIGION

Religion has prophecies regarding the end of the world. *And as [Jesus] sat upon the mount of Olives, the disciples came unto him privately, saying, Tell us, when shall these things be? And what shall be the sign of thy coming, and of the end of the world?*[82]

There will be earthquakes, pestilence, and wars.[83] We've already had lots of those.

The seven trumpets of the angels will sound; a third of the sun and moon will be darkened.[84] Revelation 8:8-9 speaks of one-third of the sea turning to blood, and a third of living things dying in the sea. The seven angels pour out the vials of the wrath of God upon the earth. Revelation 16:3 says the sea will turn to blood, and every living soul will die in the sea. There will be a great plague of hail.[85]

And the four angels were loosed, which were prepared for an hour, and a day, and a month, and a year, for to slay the third part of men.[86] The number of horsemen is to be two hundred thousand thousand, or 200 million.[87] Still, the rest of the men not killed by these plagues do not repent.[88]

There is a description of a beast, and all that dwell on earth worship him, unless their names are written in the book of life.[89]

My view is that things will not be quite as bad as all this sounds, unless you are on the hell list. The sounding of the seven trumpets of the angels may refer to the events not only right at the last minute, but during the Second Coming itself. For example, for every living soul to die in the sea would be quite traumatic for us today. It may not be so bad if it refers to the earth being changed at the Second Coming into a more glorious form so that there literally is no more sea.[90]

THE GOOD NEWS

There are two ways the world can come to an end: God can do it, or it will be our do-it-yourself project.

The good news of doing it ourselves is that we will never finish the job. If we have a nuclear holocaust, one-third of us may die, but the rest will survive. That is why we have dystopian movies. In the backs of our minds, we realize that we may have a global catastrophic event, but some of us will survive and be left in the aftermath or dystopia. Dystopian is the opposite of utopian. Utopian is the modern garden of Eden, a peaceful, wonderful society of enlightened people. Dystopian is a violent, chaotic world full of destroyed cities.

If you want to give yourself a headache, try to imagine our messed-up society continuing for 20,000 years. Keeping low population on a planet for six thousand years of Biblically-recorded history in largely agrarian societies wasn't hard, but technology accelerates destruction.

Dystopian movies continue to be popular in America. We sense things are getting worse in our country, and in the world, and we want to know how bad it can get. We want reassurance that we, with our friends and family, will survive even if there is war, destruction, aliens, large monsters, or a plague of zombies.

THE FUTURE

What do we tell our children and grandchildren about the future? Is the apocalypse a good time to be alive?

Mortal life is designed to give us eighty years of experience with good and evil, as a preparation for fourteen billion years of peace. We feel turmoil in our lives. We live in America, the most blessed country in thousands of years, the pinnacle of what God wants for his children. Many others throughout the world live with government oppression, violence, and tyranny.

The point of the frightening descriptions in Revelation isn't to scare all the good people to death. The point is that the bullies of the earth are going to be overthrown. Looking at the nations of the earth, how many are ruled by persons, groups, and regimes whose primary purpose is their own power and benefit, with the welfare of the populace being secondary? Modern nations and regimes continue to engage in warfare for evil purposes. Genocide and ethnic cleansing are common.

The vast majority of people in these countries are the victims.

We have a world full of good people who are oppressed because mortals are doing a poor job of governing themselves.

When it comes down to it, is all the suffering worth it?

There is no better test of the question than looking at the apocalypse, the end of the world.

For then shall be great tribulation, such as was not since the beginning of the world to this time, no, nor ever shall be. And except those days should be shortened, there should no flesh be saved: but for the elect's sake those days shall be shortened.[91]

What I say to all of us, and to our children and grandchildren, is that it is worth it. God does love us. Living through the apocalypse is worth it.

The reason is that mortal life is magnificent. We should not be judging our own lives by our fears. We might get a divorce. We might go bankrupt. We might die of cancer, or worse, our child might die of cancer. The apocalypse is coming, literally. What I see in the generation of people I know who have faced death, my parents and grandparents, is that they lived full lives. They accomplished in their lifetimes what God wanted them to. They were good, honest people. I love them for that. I stand and salute them all, because every last one has now given his or her mortal life for this cause, the cause of BEING MORTAL: the cause of living with evil; and the cause of enduring the constant spiritual pain that accompanies every eternal, gracious, beautiful, kind, loving, and spiritual being who dares to choose to be mortal. Our loved ones who have died have all accepted death and the grave. They lived trying to obey God, and they died hoping he would wipe the tears from their eyes. The evil we face need not cause us constant fear and anxiety. We need not let our lives be ruled by fear.

We know we are facing the greatest evil the world has known. So what if we take anti-anxiety medications?

Our parents and grandparents lived rich, full lives. They lived through the great depression and World War II. They also lived through decades of America the Wonderful, the great city and nation-state of eternal freedom dwelling amidst mortals, defying evil.

The truth and the reality is that we are standing fearlessly. We do commit adultery. We do lie, cheat, and steal. We defy God often and stoically. We become atheists. If God wants to hide, let him. We are independent beings, and God is letting us be independent. God designed this planet to allow us to choose. He knew it would be worth it to us to have a chance to do evil, to see how it feels, and then to repent. We all will repent, almost every last one of us, all except the sons of perdition.

You have a conscience you cannot get rid of, because it is

your mind and your spirit. Your mind and your spirit are within you, more powerful and lasting than your mortal body. God is calling your name through your spirit. He is telling you that it is better to be faithful than to commit adultery. He is telling you it is better to clean up and straighten up. It is better to love than to quarrel and hate.

When we die, I hope you and I get to meet the being of light. I think every last one of us deserves that. You deserve to have the being of light ask you what you've done with your life. Was it worth it? Did it make you happy? You deserve to weep with grief, as you say: "No, I wasn't happy. I tried so hard! I loved God so much! All I ever wanted was to make him happy. But I was so weak. I was so frail. My whole life feels like an endless stream of mistakes. I hurt the people I loved more than life itself. I didn't want to, okay I did want to, but I also didn't , don't you see? Don't you see the conflict within me, the weakness? Don't you understand how much I wanted to see the face of God? Don't you know I felt abandoned by God? Where was he? The thing I wanted more than anything in the mortal world was just once to look in the face of God, and know that he lives, and see love in his eyes. I want to see that he loves me. Please."

If you meet the being of light, you will see your life. You will see it because it is recorded in your soul, your eternal self. You will know that the being of light is a giver, not the judge. You will see the gift of mortal life, your mortal life. You probably weren't a president, senator, rock star, or movie star. You probably weren't rich and famous. Very few are. If you were, those experiences are part of your life, and you will see how they helped you grow, just as we all will see how our life's experiences have helped us grow.

We have become.

We are the fathers, the mothers, the grandfathers and

grandmothers, the sons and daughters who chose to dwell as mortals on earth. We are the kind neighbors. We ran the charities, and fed the hungry. We gave shelter to the homeless. We gave our lives in service, taking our children to school every morning, or to soccer practice. We are the ones who loved animals, because we recognized that they are spirits, they are minds, within physical animal bodies. We delighted in their spirits, their personalities, and their innocence.

You will see your life then as a completed thing, a task, your own mission, and one that at times seemed impossible, but wasn't, only because God helped you. You will see eighty Christmases, and Fourths of July, and/or any other holidays you learned to love in life. You will see a career of service, or remember being a stay-at-home-mom, or dad, or whatever you had the privilege of becoming. You will see the experience was so much more good than evil. You can count the beautiful spring mornings, the green grass and trees, the life and the butterflies you so often overlooked as a mortal, but which were being recorded in your soul to eternally overwhelm whatever evil you saw.

You will see then that the being of light has always let you choose. He kept you free. You are like a beautiful bird, and he set you free. He let you fly. He gave you a whole world to fly in. And he guaranteed that if anything bad happens to you, he will save you, and let you fly eternally, in exactly the kind of world you choose, a world beautiful beyond what you imagine, a reality better than your fondest dreams.

HELL IS OUR OWN CHOICE MORE THAN IT IS GOD'S PUNISHMENT

I'm hoping everyone gets to meet the being of light. I think everyone has had to live through hell on earth, to some degree, and we all deserve to meet the being of light.

Let's suppose that you are on the hell list. This isn't for people who committed adultery once and then spent the rest of their life regretting it, even though they feel forgiven. The hell list is for people who are liars, thieves, and murderers when they die. It isn't for all of us who will still be sinning when we die, because we all will be. We might even tell a lie right before we die, but death-bed lying doesn't count. It doesn't get us sent to hell.

Being on the hell list does get you sent to hell, eventually. It doesn't happen when you die, it happens at the Second Coming. Timing is everything.

Being on the hell list is reserved for those who become the kind of person who would rather take than give.

We are all in spiritual pain as mortals. The more sensitive we are, the more sensitive our souls are, the more pain we feel. We make mistakes. We do mean and ugly things. We steal. We feel terrible. We do more bad things. Our mortal lives go to hell. Through it all, those who aren't on the hell list keep trying to get better.

Those on the hell list don't get better. Some get so mad they never truly feel remorse, even to the day they die. They die with taking in their hearts, not giving. If they have a choice to give a dollar or take a dollar, they'll take. If they can make a fortune, or give one away, they'll make.

We become worse because of fear and anxiety. We look around at all the poor among us, the weak, and the suckers; they are everywhere. There is never enough to feed all those poor saps, we think. If we are the poor, we know there isn't enough to pay the rent. There isn't enough to feed our kids sometimes.

We all see the poverty and problems around us. What kind of people do we become? A great many poor people have died as givers. They died as Americans. They loved freedom, and they used their impoverished freedom to give, to lift, and to help. We aren't saying they never had drug-addicted children, but we're looking at what they did about it. They chose to give. They helped raise their own drug-addicted children, and the babies of those children. They gave the best food they had to those grandchildren. They bought them clothes for school, and a dress for church on Sunday.

What happens when you die, if you are on the hell list?

I hope you meet the being of light. You will see your life, whether in a flash, or for the next thousand years, or both, because your life is recorded in your soul.

You are going to see those givers, the impoverished Americans, the free, the Americans in spirit, the beings from the eternal world, who lived mortality by becoming givers. They gave in spite of their poverty. They did not become thieves and murderers. They lifted, raised and gave, in the ghettos and slums of America and the world. They weren't all women either. There are plenty of men out there giving, lifting, sharing, and buying shoes, dresses, and basketballs for their kids and grandkids. They are paying for schooling, and helping out with college.

Those are the Americans I know.

So if you are on the hell list, get off. Change. Become a giver, not a taker. Give love and shoes, not violence and excuses. Otherwise, when you die, and meet the being of light, and you will, either at the Second Coming or sooner, you will want to run and hide. You will want to crawl in a cave or have a mountain fall on you. You will know the Americans, the eternal free, were giving, while you were taking. You lived in the same neighborhood, but you were shooting. They were standing in front of children, using their mortal bodies as shields against your bullets.

WHAT HAPPENS NEXT

Unless you die before the Second Coming of Christ, you still have to get through the apocalypse.

We want to know what comes next, for us as individuals, for America, and for the world.

Here is what happens next: You, me, our children, and our grandchildren are going to fight like Americans to save not just America, but the world.

God loves us, because we are his kind of people. He is on our side. The Book of Revelation written by John was not given by God to scare us or our children. It was given to encourage us.

The Book of Revelation shows that God loves America and freedom, and he is going to establish freedom forever.

Tyrants and thugs should be scared. They are going to meet the being of light, within the next few decades, if they die normally. Whether the Second Coming is in a decade or a century, those who are on the hell list are going to die both mortally and spiritually at the Second Coming. Spiritual death means God is not going to be available to them for a thousand years. No matter how much they would like to reach out to him and feel his forgiveness, it will be too late. The Book of Revelation is a warning for tyrants, liars, thieves, and murderers. For tyrants and oppressive regimes, the end of their power is coming, and a thousand years is a long time to be in hell.

Let those who face hell be worried about a third of men dying on earth, or about great plagues.

For the rest of us, we are looking forward to the Second Coming, in the sense that we will get to see and be with Christ, whether we are mortal or in the spirit world.

Until then, we have the assurance of God's love and protection for all those who are fighting for freedom, for all who are choosing good over evil, and for all who live daily with hope.

THE GREATEST GENERATION

We rightly view those who lived and fought through World War II as a great generation of Americans. A greater generation exists in our youth today. They stand with a bright hope in spite of the challenges in our society. They are capable of meeting every coming challenge. They are anxious to continue inventing new science, making new discoveries, and finding new cures.

The future looks bright for America. I'm not saying that because we won't face the apocalypse. We will. We already know that both good and evil people have died and do die in plagues and earthquakes. The difference is that those who stay off the hell list by becoming givers, not takers, will face the being of light for consolation. The being of light effectively wipes the tears from their eyes by showing them that they are not condemned to hell for having been weak, sinful mortals. Those on the hell list are going to see the example of those who kept trying and giving, no matter how bad it got. Those on the hell list are going to feel rotten not because God is vindictive, but because of their own common sense and conscience, knowing they had a choice to be givers, not takers, and they chose wrong.

A SHORT LIST OF CATASTROPHES

The best thing that can happen is that we won't have any catastrophes to undermine our freedom in America before Christ comes. Granted, this isn't likely.

The worst things that can happen are:

FINANCIAL DECLINE OR FINANCIAL MELTDOWN

We have seen financial decline before. We know about the great depression in the 1930's. We remember the mortgage crisis and recession from 2008.

Increasing national debt in America means we are paying more interest. Having our taxes increase because we have to pay so much interest is painful.

We've seen hundreds of years of economic growth because our population was growing, and we had an abundance of resources.

We face a decline. Our children may be worse off financially than we are.

The problem isn't that America doesn't have enough food to feed our population. We have an abundance of housing. The future in a declining economy doesn't hold mass starvation physically, it holds spiritual challenge—the discouragement that comes from lack of purpose.

There are simple solutions: Suppose an increasing percentage of millennials live with their parents. Households include three generations, grandparents, parents, and children. We shift to public transportation because we can't afford cars.

In short, we get practical, but we don't starve. The key is to not get discouraged. The truth is that we don't all need large homes and expensive cars. We could get by on a lot less. We could be happy, even if we had less. I'm pretty sure we are going to have the chance to find out, and that is okay. We'll be givers, not takers.

DEMOGRAPHIC SHIFTS

Nations and populations may be self-destructing by not having children. A number of books have been written about it.[92]

LOSS OF THE POWER GRID: We lose electrical power, and it takes several years to rebuild our systems.[93] The key issues if this happens will be whether we maintain law and order, food and water supplies, and medical care. We are highly vulnerable to this disaster.

EMP STRIKE: Electromagnetic pulses could wipe out our transportation and vehicles, as well as our power grid. The issues are whether the military and police will be able to mobilize and maintain law and order, whether we have access to food and water, and whether we can receive medical care.

NUCLEAR STRIKE: A nuclear strike adds the complication of radiation contamination. A single strike on Washington could eliminate a large percentage of our nation's leaders. Multiple strikes, or hundreds of strikes, could be catastrophic.

PLAGUE: It is possible for a disease to skip the bounds of containment and cause massive casualties. We saw the potential with the ebola epidemic in Africa. It was eventually contained. The rates of casualties in the worst-affected areas signal what could happen worldwide.

NATURAL DISASTER: A meteor strike could cause catastrophe. The usual earthquakes, tidal waves and floods have not posed catastrophic threats. Though they have devastated isolated areas, they haven't threatened our nation's functioning.

INVASION: Invasion of America by foreign powers is always possible. The conquering of America would extinguish the beacon of freedom in the world.

ALIENS FROM OUTER SPACE AND ZOMBIES: There is no such thing as a little green alien from outer space. The same goes for zombies.

AMERICA THE FREE

In each disaster scenario, the issues can be stated simply: Does the United States remain as a free country? What percentage of our population dies? Does our government function and communicate with us? Do we maintain law and order through police and military? Do we have sufficient food and water distribution from farms or storage to prevent starvation?

I am optimistic because God loves people who are trying their hardest, even though they are still sinners. I am optimistic because I see a difference between most people and those on the hell list. Even for those on the hell list, I still have hope. There is still time to be kinder, better, and to become a giver.

WHAT TO DO

The answer seems deceptively simple:

Get off and stay off the hell list.

Work as hard as you can serving others; become a giver.

Get as much education as you can.

Be a kind and good person.

Stop being afraid or anxious.

Stop being contentious.

STORE PROVISIONS

Learn what it takes to store and rotate food supplies, water, and other essentials.

GET A COLLEGE DEGREE

Pursue college and post-graduate studies to improve your economic outlook. Education has value of itself, even if it doesn't lead directly to a better job. There is even some consolation if the end of the world comes: Your student loans would be wiped out. Odds are that you have plenty of time, if you are a young person or young adult, to have a great future, with or without the Second Coming.

THE END OF THE WORLD

The trigger event for the end of the world is abandoning law and freedom, and turning to violence. If Americans, the spiritually free, are in danger of extermination, God will save them.

The United States of America is God's government of world leadership for freedom.

Dystopian society would begin if the United States fell. No other country could or would replace it. God never intended for Americans, those who love and fight for freedom, to be wiped out. He fully expects us to continue to fight for freedom, and he will intercede before he ever allows us to be destroyed.

ARMAGEDDON

The other trigger event for the end of the world is a siege on Jerusalem.[94]

If you fear the great battle of Armageddon, keep in mind who wins.[95]

THE GREAT SOLUTION

We are not exactly in control of world events.

We do have some control, and we need to do all we can to improve our lives.

As a society, America has been the great solution. The

solution, simply put, was the freedom to go from the Revolutionary War to becoming the greatest economic and military superpower in the history of the world.

Now our country is paralyzed by national debt. We sold our birthright of freedom by spending our children's inheritance, leaving them instead with a debt that neither we nor they can pay. God knows that we are fallible humans, and that even in America we would eventually let freedom slip from our grasp. It is a flawed mortal system. We're still the best country this world has ever known, even when we are staring down the greatest evils the world has ever known. Especially then.

We are at the zenith of world society. America did not do it alone. America's allies, those who have loved and fought for freedom anywhere and everywhere in the world, have fought with us.

We have felt to mourn that America will decline, and for years we have feared she will one day be lost, and we will be left in a world of chaos, violence, and tyranny.

What we have not understood is that God timed the events this way. He knew the day would come when our grasp on freedom would slip.

What God sees, which we may not have seen until now, is that America succeeded.

AMERICA HAS DONE WHAT GOD ASKED HER TO DO

We as Americans, from our founding fathers and mothers to our children, have done the work God gave us to do.

We have already won! We have already preserved freedom in the world for longer than might have been expected. Even though we now face our greatest challenges, we have not and will not fail.

Can God save Americans and all their allies in freedom, along with every person who has lived and died for freedom, family, and country, since the dawn of this great nation and before?

He has, he can, and he will.

God timed the events so that we would have the freedom we needed between 1776 and today. I believe we have many decades left, but if Christ's Second Coming were tomorrow, we have already accomplished what God put us here for. It is up to us to keep and maintain our country and our freedom in decades to come.

We are the solution. If the day comes when we can no longer maintain our freedom, the great solution is the Second Coming of Jesus Christ. When it comes, all of us who have stood, and are standing even now, and always will stand for the cause of freedom, will greet the one who brings freedom.

This great nation, the United States of America, is an integral part of the stone which was cut without hands, seen by Daniel. That stone is nothing less than the collective lives, sweat, and blood of every person who has fought for God and for freedom. Some fought in the reformation. Some fought in the Renaissance. Some inspired us with the greatest music ever written. Some were the greatest scientists in the history of the world. Some were inspired with the technologies that saved us in war after war, through our most troubled and darkest hours, when the torches of freedom flickered in pervasive darkness.

We await Christ. We fight while we wait, and we know we cannot and will not lose. He did not set everything happening in the world right now just to have the best people he has on earth all destroyed before he comes again.

WHAT WE ARE BECOMING, INDIVIDUALLY AND AS A SOCIETY

The good news when God ends the world is that he makes the world better. Those who remain are going to live in a world with a paradigm shift: Animals will stop eating each other, and even humans will start being nice.[96] The lamb and the lion shall lie down together; in other words, we will realize the lamb and the lion are the same, and they are Christ: Christ was the lamb when he was slain. He returns as the lion, and we see that the two sides of his nature are united. He is the strongest of all, yet he willingly died for all.

That's what is happening to each of us: we are following the pattern of Christ from lamb to lion by living in a world with good and evil. We start out as babes, as little lambs, innocent from birth, and we grow to become strong and powerful. That is the purpose of our confrontation with evil. It is the reason for our duality, and for the constant conflict within us. Divorce, cancer, and natural disasters are among the many evils we may face in life. We could not grow in power without the evil in the world. We could not become stronger. God would not exalt us as ignoramuses. He wanted us to know and understand how bad things can be when there is no strength. Someone better have enough strength, power, armies, or technology to defeat those who choose evil. God cast out Satan because God was stronger. He knew more. He had better technology—in fact he perfectly understands all science and technology, not just on our little planet with its nukes and supercomputers, but throughout his cosmos, amidst his galaxies.

We need to have the experience of enduring an evil world. We are not spared from cancer, divorce and bankruptcy. These things are inflicted on us with all the precision of an infinite computer. We sometimes feel we have been pushed to the utmost limits of our capacity. God, or life, may force us to do things which are extremely painful for us. We may have to grow in ways we don't think possible.

There is a great schism between our dreams and reality. Our experience becomes reality. The joy and suffering of eighty years in mortality and the experience of being spirits is the warm-up for an eternity. The reality of our eternal state exceeds our dreams. The reality that God saves all but the sons of perdition, in an eternity of peace, is beyond our fondest imagination.

There is a great schism between our American Dream and reality. We see a glimpse of the reality that God set up America as the epitome of what a nation of strong mortals can do if they love freedom, family, and God. We, the people of freedom, have ruled this nation.

All of those who love and fight for freedom have helped roll the stone, the Kingdom of God as seen by Daniel, a process that began in 30 A.D., and reached a zenith in the garden of good and evil in America in our generation.

America is the precursor nation. After we have done all we can do, and before we are ever destroyed, Christ will come, with a nation for eternity.

America has exceeded our dream.

CHAPTER ELEVEN

ETERNITY: THE FATE OF THE UNIVERSE

SPECULATIVE JUDGMENT

I speculate that a lot more people are going to get transfigured, and caught up into the air, at Judgment Day, than most of us realize.

In my view, Catholics, Protestants, Hindus, and atheists will all be transfigured, regardless of religion, unless they are on the hell-list.

Most of the people I know are good people. I don't mean they are perfect. They sometimes lie, exaggerate, are selfish, and are mean. I have a hard time playing a board game without getting mad at someone. On bad days, I cuss in traffic.

Most of the people I know, including me, feel guilty. We feel guilty when we are selfish and stupid. We even feel guilty when we don't measure up. We have too much credit card debt. We sometimes buy stuff we don't need. We feel guilty when we are grouchy and in a bad mood, but sometimes we are in a bad mood anyway.

Most of us are struggling, but we are good people. We recognize right from wrong. To me, that is the standard.

I speculate that people who are professional thieves, or who have committed murder, will be burned at Judgment Day, and will go to hell for a thousand years. Members of drug cartels, and murderous gangs, seem likely candidates for hell. There have been more than a few of these people.

The rule of law in the U.S. makes it hard to be a liar, thief, or murderer, and get by. Most of us have to go to work every day. Most of us recognize the importance of community service, and helping others.

Religion is clear on how important it is to preach repentance. There is a good reason for that, and it isn't just religion. With or without God, most of us want our children to stay away from illegal drugs, work hard, and to go to college. We don't like our children to lie, or steal. As parents, we preach a lot of repentance. We teach our children to be good people.

We should encourage, as a society, sexual abstinence outside of marriage. It is a moral issue. It is important.

I'm saying people should straighten up. I am telling people they should be worried about things we know, in our consciences and hearts, are wrong. We should be trying to become kinder, better, and more honest people.

TRANSFIGURED: THE NEXT PARADIGM

The vast majority of people who have lived on this planet did not become liars, thieves, and murderers. They did not stop listening to their consciences.

The majority of people, a speculative eighty billion minus the hellions, are going to be transfigured, and caught up in the air, technologically, to meet Christ. Mortals will be transfigured. At least temporarily, they will not be mortal. Those who have died before, and gone to the spirit world or paradise, may be resurrected.

The earth will melt with fervent heat, and become a transfigured planet. It will be technologically altered. It will return to a Garden of Eden technology: fruit to eat, and no weeds.

The earth will become heaven, or a technological utopia, if you prefer, with Christ present among those who wish to be with him.

SATAN AND HIS FOLLOWERS WILL BE LEFT WITHOUT A PLANET

For a thousand years, Satan and his followers, and those liars, thieves and murderers who stopped listening to their consciences, will be without a planet. I think the earth will be moved to another location, and those in hell will remain in this solar system without the earth here. Sure, they can go to Mars, but they will be unable to build anything, any more than Satan and his followers can build anything in a mortal world. They are technologically incapable. Satan won't be able to build so much as a tin shed, on Mars, or anywhere else.

HELL FOR EX-MORTALS

The liars, thieves, and murderers who go to hell are going to be left for a thousand years with the technological nakedness of their own consciences: a perfect memory of what they did wrong, and of the suffering they caused others, and no ability to repent, until the thousand years ends.

In addition to being unable to regain a clear conscience for a thousand years, those going to hell at Judgment Day will be cast out with Satan and his followers. Hell is physically defined as being without a planet. It is referred to as being cast into a pit, perhaps the symbolical hole left in our solar system where the earth used to be.

HEAVEN AND HELL GO DIGITAL

Heaven and hell have gone digital. Heaven and hell are technological paradigms. They are executable programs. Not believing in God isn't going to stop the program. Heaven can be technologically defined as a transfigured state, with both the planet and the people on it being transfigured, using Type D technology. Resurrected beings, Type C technology, will be among the transfigured ones. All will look human.

The experience of two men on the road to Emmaus tells us the options of this technological utopia. They didn't see the resurrected Jesus Christ as an angel in a starkly noticeable, shining white robe. They saw him as another person, wearing normal clothes.

Transfigured beings can return to mortality; Moses, after being transfigured and talking to God on the mount, came back, veiled his face temporarily, and continued life as a mortal once his face stopped shining. In the alternative, a person can stay transfigured for hundreds of years, as Elijah did. He left in a chariot in about 900 B.C., and came back in 33 A.D. on the mount of transfiguration.[97]

Hell is technologically defined as being stuck as a spirit matter person, without a nice planet to live on. It means being unable to travel faster than the speed of light. Spirits aren't made of atoms or photons, and therefore wouldn't necessarily be subject to the speed of light; I think they are anyway, for other reasons. For example, it would take twenty seven thousand years, if a spirit/Type A matter person could travel at the speed of light, for him or her to get to the center of our galaxy. He would be better off to just stay put. Four years would get him to the nearest star, which probably isn't any better than the one we already have. Oops, don't get lost. Or go crazy, out there alone. They can't travel fast enough to get to anywhere important, which is the same as being imprisoned, or cast out. It's going to be hell.

MOVING PLANETS

There is reason to think that the transfigured earth is going to be moved to a different location: *And the stars of heaven fell unto the earth, even as a fig tree casteth her untimely figs, when she is shaken of a mighty wind.*[98] Moving a planet is high-tech.

FOUR PARADIGMS

If there are four types of beings, there are possibly four types of planets.

After a thousand years, earth gets upgraded from transfigured technology to resurrected technology. Now, that's heaven.

This earth started as a spirit matter planet, got upgraded to a mortal planet, and will be upgraded to a transfigured planet. At the end, it will be a resurrected planet, the sea of glass and fire John the Revelator saw. That is technological power, not mythology or magic.

EVOLUTION

We are evolving into eternal beings. It is small-minded to suggest that humans are going to stay mortal, on a mortal planet, without access to higher technology. Mortal matter is so wonderful; we have a whole planet, and we can run it, and we have lots of physical mortal joy and pain. We can screw everything up, and make all kinds of mistakes, and eventually, end up with utopia anyway.

Utopia comes from <u>not</u> being mortal. Mortal is a wonderful design, but the best use of this technology is for experimental, experiential, planets. It is for a temporary field trip, a theme park, where accidents and bad decisions, and sins, can be repaired, eventually, perfectly.

Atheistic scientists, and Hindus, and Christians, are all evolving, and they will all be judged, at Judgment Day, by the standard of conscience, with technological, better-than-computer precision. Yes, it is religious, but it works just as effectively technologically.

After judgment, we evolve to other paradigms of technology. The difference between heaven and secular utopia is whether you want to accept God's offer of religion, which is the same as accepting a relationship to him: accepting him as God. If you don't want to, you are still going to be in utopia. That is freedom. That is the opposite of a condemnation mindset. Condemnation is for willfully becoming, as a mortal, a liar, thief, or murderer, to the point where your conscience is numb, and you have no desire to change. Hell is for that. Hell is a punishment for that.

Remember Thomas. He didn't believe Jesus Christ was resurrected until he saw. That's a lot like an atheist. But he went to heaven anyway. More blessed were those who believed, yes, and that is its own reward and joy.

If you don't accept religion, you are still evolving, and not into an eternally miserable, suffering person. You will evolve into a resurrected being, with or without belief in God. He'll give you that, regardless of what you believed.

THE ETERNAL UNIVERSE

God has the technology for this universe to continue indefinitely, eternally. We will fit in the order of things, as ex-mortals. Either we will be with God in an eternal heaven, or this earth as John the Revelator saw it, which will be a resurrected, technologically-upgraded planet, or we will live as resurrected beings on a planet like this earth without the upgrade, a utopia, without God.

THE GREAT, FLOWING CURRENT OF KNOWLEDGE

We know about the internet. It operates and functions because of electricity, photons, light, and electromagnetic forces which can carry, record, and communicate data, which is knowledge.

Everything is data. Everything is digital, but it isn't the mortal internet we are talking about. There are other types of light, not made of photons, more glorious, more powerful, and technologically superior. No scientist can say that the speed limit of mortal light (photons) applies to more advanced types of light.

Prayers are technology. Answers to prayers are technology.

There is likely a vast database of digitized information, containing all data in the universe. There is ongoing communication, the great internet, not just to our minds, from God, but also to animals. From atoms to galaxies, everything acts like it is listening to instructions, as if plugged into, and connected to, a vast internet. The laws we understand, gravity, electromagnetic, strong and weak forces, are all operating with startling, flawless, digital precision.

Omnipotence is to have perfected technology, and after fourteen billion years, God has.

Omniscience is being smart enough to record everything, technologically, and to be able to access the database. It means overcoming the issue of time, and accessing past, present, and future.

What would you do, if you were God? Exactly what he has done, and for the same reasons.

Mark Hawkins was born and raised in Cody, Wyoming. He obtained a bachelor's degree in Business Management—Finance, and completed his Juris Doctorate in 1990. Visit the website at markwhawkins.com

ENDNOTES

[1] Exodus 20:2-3. All Biblical references are to the King James version.

[2] Stephen W. Hawking, *The Theory of Everything: The Origin and Fate of the Universe*, Jaico, Fort, Mumbai, 2009.

[3] For further reading: Lisa Randall, *Knocking on Heaven's Door*, HarperCollins, New York, 2012. Lisa Randall, *Warped Passages: Unraveling the Mysteries of the Universe's Hidden Dimensions*, HarperCollins, New York, 2006.

[4] Daniel 2:1-49.

[5] Genesis 37:9-11.

[6] Genesis 40:1-23.

[7] Genesis 41, 42.

[8] Exodus 28:30; Leviticus 8:8; and Numbers 27:21.

[9] 1 Samuel 9:9.

[10] Acts 2:1-13.

[11] Deuteronomy 34:5-6; Mark 9:2-10; and 2 Peter 1:16-21.

[12] Daniel 3:1-30.

[13] Romans 8:16-17.

[14] Revelation 15:2.

[15] Genesis 5:5.

[16] John 11:43-44.

[17] Luke 24:36-43.

[18] Revelation 12:9.

[19] Luke 9:28-31.

[20] Luke 9:29.

[21] Exodus 34:28-35.

[22] 1 Samuel 28:15-19.

[23] For further reading: Rabbi Joseph Telushkin, *Jewish Literacy*, New York, HarperCollins Publishers, 2001; and Ed. Arthur Herzberg, *Judaism*, New York, George Braziller, Inc., 1961 .

[2424] Revelation 21:8.

[25] Revelation 21:8.

[26] Luke 24:37-39.

[27] Luke 1:19.

[28] Luke 24:36-43.

[29] 1 Peter 4:6.

[30] Revelation 20:4.

[31] Luke 24:42-43.

[32] Revelation 4:6; 15:2; and 21:18, 21.

[33] 1 Corinthians 15:22.

[34] Philippians 2:10

[35] Exodus 25:22.

[36] Ezekiel 1:4.

[37] Revelation 21:2-3, 10; Revelation 4:2-10; Exodus 25:22; Isaiah 6:1-2; and Ezekiel 1:9.

[38] Revelation 4:1 and 21:2-3, 10.

[39] James 2:26.

[40] Luke 1:19.

[41] For further reading: Lewis, Thomas. *The Lives of a Cell: Notes of a Biology Watcher*, Bantam Books, Inc., 1974, pp. 12-13, 18.

[42] Exodus 25:22.

[43] Matthew 15:32-38; John 2:1-11.

[44] Exodus 25:22.

[45] Luke 1:19.

[46] Matthew 7:7.

[47] Luke 23:43.

[48] Luke 24:6-7.

[49] John 20:17; Revelation 2:7; and Revelation 22:1-2.

[50] For further reading: Lutzer, Erwin W., *One Minute After You Die*, Moody Puplishers, Chicago, 1997.

[51] Revelation 12:4, 9.

[52] Jude 1:6; Revelation 12:9.

[53] Revelation 12:4, 9;

[54] Genesis 4:8.

[55] Revelation 20:1-5.

[56] James 2:26.

[57] Matthew 4:1-11; 1 Corinthians 7:5.

[58] Revelation 20:8.

[59] Malachi 3:5.

[60] Psalms 16:10; Acts 2:25-27; and Revelation 20:5.

[61] Revelation 20:5.

[62] Revelation 20:2-5.

[63] Acts 2:25-27.

[64] Revelation 20:1-5.

[65] Revelation 20:7-9.

[66] Romans 8: 16.

[67] *127 Hours*, Fox Searchlight Pictures, Directed by Danny Boyle (2010).

[68] Malachi 3:2.

[69] *America the Beautiful*, Katherine Lee Bates, 1859-1929.

[70] Daniel, Chapter 2:31-34.

[71] For Further Reading: "Daniel's Stone that Became a Mountain", Articles by Watchman Radio Hour with Alex Dodson, www.oneplace.com/ministries/watchman-radio-hour/read/articles/daniels-stone-that-became-a-mountain-11790.html.

[72] Daniel 2:34-35.

[73] The Constitution of the United States of America.

[74] Luke 16:26.

[75] Matthew 27:50-51.

[76] 1 Peter 4:6.

[77] Luke 24:39.

[78] 1 Peter 3:8.

[79] For reading about networks: Ramo, Joshua Cooper, *The Seventh Sense: Power, Fortune, and Survival in the Age of Networks,* Little, Brown and Company, 2016.

[80] 2 Corinthians 3:17.

[81] Philippians 2:10-11; John 17:12.

[82] Matthew 24:3.

[83] Matthew 24:6-8.

[84] Revelation 8.

[85] Revelation 16:21.

[86] Revelation 9:15-18.

[87] Revelation 9:16.

[88] Revelation 9:20-21.

[89] Revelation 13, especially v. 8.

[90] Isaiah 50:2.

[91] Matthew 24:21-22.

[92] For further reading: Buchanon, Patrick J., *The Death of the West: How Dying Populations and Immigrant Invasions Imperil our Country and Civilization,* St. Martin's Press, New York, 2002; Steyn, Mark, *America Alone: The End of the World as We Know It,* Regnery Publishing, Inc., Washington, D.C., 2006; Steyn, Mark, *After America: Get Ready for Armageddon,* Regnery Publishing, Inc., Washington, D.C., 2011; Goldman, David P., *It's Not the End of the World: It's Just the End of You: The Great Extinction of the Nations,* RVP Publishers Inc., New York, 2011; Goldman, David P., *How Civilizations Die (And Why Islam is Dying Too),* Regnery Publishing, Inc., Washington, D.C., 2011.

[93] For further reading: Koppel, Ted, *Lights Out,* Crown Publishers, 2015.

[94] Zechariah 12-14; Matthew 24:15-22; and Revelation 16:14-21.

[95] Zechariah 12:2-3, 9.

[96] *and the calf and the young lion and the fatling together…* Isaiah 11:6.

[97] 2 Kings 2:11-12; and Matthew 17:1-9.

[98] Revelation 6:13.

Made in United States
North Haven, CT
04 November 2026

21765277R00104